Michel Serres and French Philosophy of Science

Michel Serres and Material Futures

Series Editors: Joanna Hodge and David Webb

Series Editorial Board: Claire Colebrook, Steve Connor, Diane Morgan, Dan Smith, Iris van der Tuin, Chris Watkin

Serres is a radically disruptive thinker with respect to the history of philosophy, the practice of philosophy and the future of enquiry. His writings stretch from the 1960s to the present moment.

This book series provides discussions of both of the various stages and aspects of his own writings, and extend the discussion to developing responses to contemporary questions and problems of and for philosophy including mathematics, our relationship with technology and the ecological crisis.

Titles published in the series:

Mathematics and Information in the Philosophy of Michel Serres, by Vera Bühlmann
Michel Serres and the Crises of the Contemporary, edited by Rick Dolphijn
Michel Serres and French Philosophy of Science, by Massimiliano Simons

Michel Serres and French Philosophy of Science

Materiality, Ecology and Quasi-Objects

Massimiliano Simons

BLOOMSBURY ACADEMIC
LONDON • NEW YORK • OXFORD • NEW DELHI • SYDNEY

BLOOMSBURY ACADEMIC
Bloomsbury Publishing Plc
50 Bedford Square, London, WC1B 3DP, UK
1385 Broadway, New York, NY 10018, USA
29 Earlsfort Terrace, Dublin 2, Ireland

BLOOMSBURY, BLOOMSBURY ACADEMIC and the Diana logo are trademarks of Bloomsbury Publishing Plc

First published in Great Britain 2022
This paperback edition published 2023

Copyright © Massimiliano Simons, 2022

Massimiliano Simons has asserted his right under the Copyright, Designs and Patents Act, 1988, to be identified as Author of this work.

For legal purposes the Acknowledgements on p. vi constitute an extension of this copyright page.

Series design by Irene Martinez Costa

All rights reserved. No part of this publication may be reproduced or transmitted in any form or by any means, electronic or mechanical, including photocopying, recording, or any information storage or retrieval system, without prior permission in writing from the publishers.

Bloomsbury Publishing Plc does not have any control over, or responsibility for, any third-party websites referred to or in this book. All internet addresses given in this book were correct at the time of going to press. The author and publisher regret any inconvenience caused if addresses have changed or sites have ceased to exist, but can accept no responsibility for any such changes.

A catalogue record for this book is available from the British Library.

A catalog record for this book is available from the Library of Congress.

ISBN: HB: 978-1-3502-4786-4
PB: 978-1-3502-4790-1
ePDF: 978-1-3502-4787-1
eBook: 978-1-3502-4788-8

Series: Michel Serres and Material Futures

Typeset by Deanta Global Publishing Services, Chennai, India

To find out more about our authors and books visit www.bloomsbury.com and sign up for our newsletters.

Contents

Acknowledgements		vi
Introduction		1
1	Surrationalism after Bachelard: Michel Serres and *le nouveau nouvel esprit scientifique*	13
2	Michel Serres and the epistemological break	35
3	Purification as a practice of the self	61
4	French object-oriented philosophy in the 1970s: Serres, Dagognet and Latour	87
5	Brewers of time: Michel Serres and modernity	115
6	Thanatocracy and the anthropology of science	141
7	The secularization of science	171
8	The parliament of things and the Anthropocene: How to listen to quasi-objects	197
Notes		221
Bibliography		224
Index		240

Acknowledgements

As many before me, I discovered Michel Serres's work through his book *Le contrat naturel* (1990), which remains one of the books that has profoundly shaped my thought. I have Jan Vanlommel to thank for pointing me at the book when we were still students. Similarly, it was Rudi Laermans who allowed me to write a thesis on the role of nature in sociology, thereby also forcing me to open a dialogue with Bruno Latour. But it was Paul Cortois who helped me to make sense of how Serres's work was deeply embedded in the tradition of French epistemology. Though once a side-project of my PhD, with this book I finally could put this connection at the centre.

This book could not have appeared without the helpful comments and feedback on different parts of this manuscript. I especially want to acknowledge the countless conversations with Hannes Van Engeland, as well as his detailed comments on many chapters of this book. Parts of this story were also presented at conferences throughout the years in Paris, Lisbon, Cologne, Leuven, Leeds and Hannover. Special mention should be made of the numerous interactions with the Research Network on the History and the Methods of Historical Epistemology in Paris. Also the SEP-FEP 2020 conference deserves mention, bringing Serresian scholars together in a virtual space, thereby opening up many new approaches to his work. I moreover want to thank David Webb and Joanna Hodge, who kindly supported me in writing this manuscript and for making it a part of the series on Michel Serres and Material Futures.

Four of the chapters are reworked versions of articles that were earlier published. Chapter 1 appeared as 'Surrationalism after Bachelard: Michel Serres and *le Nouveau Nouvel Esprit Scientifique*'. Chapter 2 is partly based on 'The Janus Head of Bachelard's Phenomenotechnique: From Purification to Proliferation and Back'. An earlier version of chapter 7 was published as 'Bruno Latour and the Secularization of Science'. Finally, chapter 8 has appeared in a previous version as 'The Parliament of Things and the Anthropocene: How to Listen to "Quasi-Objects"'.

Introduction

Michel Serres is an angel. I state this not to stress the moral quality of his work nor to mourn his recent death, but as a summary of his work and influence. In one of his most fascinating, yet often overlooked works, *La Légende des Anges* (1993a), Serres uses the figure of the angel as the literal legend of the map of our cultural landscape:

> Why should we be interested in angels nowadays? . . . Because our universe is organized around message-bearing systems, and because, as message-bearers, they are more numerous, complex and sophisticated than Hermes, who was only one person, and a cheat and a thief to boot. Each angel is a bearer of one or more relationships; today they exist in myriad forms, and every day we invent billions of new ones. However, we lack a philosophy of such relationships. (Serres 1993a, 293)

Indeed the philosophy of Michel Serres can be described as 'a general theory of relations' or 'a philosophy of prepositions' (Serres and Latour 1992, 127). He opposes traditional philosophies that start from the subject or the object, which neglect the third aspect of every relation: 'By that I mean the intermediary, the milieu. . . . What is between, what exists between. The middle term' (Serres 1980a, 65). For Serres these intermediary relations are the foundation of both the subject and the object (Serres 1987, 209).

But the angel is helpful to understand not only the world but also the particularities of Serres's own work, his style, and his influence. 'The good Angels pass, in silence, we forget them; the others appear and become our gods' (Serres 1993a, 104). In the same way, the role of Serres in twentieth-century philosophy is hard to pinpoint, since it tends to situate itself on this invisible plane. Hence, the difficult task of writing a book on Michel Serres.

I am not the first to recognize these difficulties, not even the first to call Serres an angel (Godin 2010, 42). Already in the preface of an early book on Serres by Anne Crahay (1988), Jean Ladrière notes how Serres's work can be read from multiple angles, similar to the work of Leibniz, not by accident the topic of Serres's dissertation:

> Michel Serres's thought is, a bit like Leibniz's system . . ., a thought with multiple points of entry. It can be approached from the perspective of the philosophy of mathematics, or that of the philosophy of nature, or that of the history of philosophy, or that of aesthetics, or that of a philosophy of communication, or in that of a reflection on history, and many more. (Ladrière 1988, 14)

But despite this apparent fruitfulness of his work, Serres never seemed to have created a large following. In the words of Christopher Watkin, we 'almost never encounter a gaggle of Serresians alongside the Deleuzians, Foucauldians and Derrideans at the larger academic conferences' (Watkin 2020, 1–2). Similarly, William Paulson, who in the 1980s had Serres 'pegged for at least as much influence and attention as was then enjoyed by Barthes, Derrida, Foucault, or Lacan,' had to confess that his prediction was 'as wrong as could be' (Paulson 2000, 215).

There are multiple hypotheses for why Serres never enjoyed the same fame. First of all, it might be related to what Watkin calls Serres's 'untimeliness' (Watkin 2020, 13). Serres was often exploring topics that would only become prominent later on, such as the role of information, objects and materiality, or the ecological crisis. However, this explanation seems unsatisfactory, since it reiterates the cliché of a genius 'ahead of his time', implying a rather linear and modern view on time and history, to which Serres himself would certainly protest (see Chapter 5).

A second hypothesis Watkin (2020, 14–15) suggests is that of Serres's interdisciplinarity, which risks either being accused of being vague, unsystematic and eclectic; or being so intimidating in the amount of presupposed erudition to make sense of his texts. In a similar vein, Latour wonders: 'Why in the space of one paragraph, do we find ourselves with the Romans then with Jules Verne then with Indo-Europeans, then, suddenly, launched with the Challenger rocket, before ending up on the bank of the Garonne river?' (Serres and Latour 1992, 43). Although more satisfactory as an explanation, one could still wonder why figures such as Jacques Lacan or Gilles Deleuze have been canonized, whose work are hardly less interdisciplinary and demanding than Serres.

A third option is suggested by Serres himself, namely that in contrast to other famous philosophers, Serres does not have one fixed concept linked to his name:

> I do not have the equivalent of this idea attached to the name of certain philosophers as if it were their treasure: the clinamen for Epicurus, the lump of wax for Descartes, the general will for Rousseau, the flesh for Merleau-Ponty,

deconstruction for Derrida, mimetism for René Girard, etc. I have no logo, no brand. But a philosophy does not spread without a logo. (Serres 2014, 360–1)

This idea is also suggested by Girard in his introduction to the English translation of Serres's *Détachement* (1983):

> Future historians of ideas may decide, at some point, that Michel Serres was one of the leading spirits in a revolution that is taking place in our midst at this very moment and is transforming our conception of knowledge. This will happen when the categories are finally created that will make his thought more predictable and classifiable than it is now. (Girard 1989, viii)

In this book I want to follow this latter suggestion. Serres's work is challenging to summarize or his influence is hard to pinpoint, because his philosophical interventions often situate themselves on an invisible, angelic plane, not easily captured by a slogan or brief summary. Instead, his work often consists of an open investigation of our most primary metaphors, distinctions and myths through which we organize the world around us.

Let me illustrate this with an example, found in *Le Cinq sens* (1985). One theme of the book is an opposition between the natural and the social sciences. Whereas the first observes, the latter criticizes and surveils (see Chapter 6). In this context Serres mobilizes an old, clichéd story: the story of the philosopher, often Thales, who with his head in the clouds fell to the bottom of a well, and is laughed at by a washerwoman. A classic tale of the unworldliness of philosophers, living in ivory towers out of touch with reality. But what does Serres do? He turns the story around, slinging it in a completely new direction:

> Who has a grasp of reality, he who gapes at the stars or she who hides in the background, making fun of him? Do the washerwomen know that a well makes an excellent telescope and that, from the bottom of this vertical cylinder, the only telescope known in Antiquity, one can see the stars in full daylight? (Serres 1985, 41)

Many of Serres's books can be seen as a quarry where old petrified stories and metaphors are excavated, examined and sent off in new directions. Another way to put it, as Christiane Frémont (2010, 19) has, is to characterize the characters that Serres mobilizes in his work – such as the Parasite, Hermes or Angels – as 'quasi-objects': entities around which a collective, a social order organizes itself (see Chapter 4). By bringing these quasi-objects into movement, one opens up radical new ways to build our whole conceptual edifice. To engage with these quasi-objects, Serres never sticks to one framework or model, but adapts it to the specificities of the phenomenon under consideration. 'I try to adapt the way in which I speak and write to the phenomena that I strive to see and grasp' (Serres 2014, 96).

Why then still write a book on Serres? William Paulson, for example, argues that though a synthetic textbook of Serresean philosophy would be a possibility, 'no one really devoted to his work could quite want to do' (Paulson 2005, 31). One would easily lose what is so fascinating in the multilayered works that Serres wrote. 'This resistance to summary or to abstract representation poses a problem to anyone who would use his writings by gleaning nuggets that can be taken away and carried over into one's own work' (Paulson 2005, 31). Watkin similarly wonders, 'how can a book that does not write like Serres begin to do justice to his thought?' (Watkin 2020, 19)

As a consequence, much of the secondary literature tends to write about Serres in a very explorative way. Such an approach is found in many of the contributions to *Mapping Michel Serres* (2005), *Le Cahier de l'Herne: Michel Serres* (2010) and most recently in the volume *Michel Serres and the Crises of the Contemporary* (2018). As Rick Dolphijn explains in his introduction, 'the contributors to the volume aim not so much to write "about" Serres ("explaining" his thoughts) as to write "with" him. In a way, their aim is to "work" with his writings; to explore how they resonate with the world' (Dolphijn 2018, 2).

Though this is a possible approach, it is not a necessary one. Such a doubt is echoed by Watkin: 'I am of the very firm conviction that a book on Serres should not need, or try, to write like Serres. To describe is not the same as to participate, and both have their place' (Watkin 2020, 21). But we need not go so far to dissociate author and commentator. What makes Serres's philosophy intriguing is not so much the lack of a rigorous style (on the contrary) nor that it is merely a mixture of all styles. Rather, it lies in the fact that all these styles are seamlessly at work next to another. Such a point of view would then imply, not an abandonment of stylistic prescriptions, but instead that every style must be taken seriously. It is not a departure from rules, styles and conventions; but a deep respect for them, though accompanied with the eternal footnote: there is no hierarchy, only a juxtaposition.

In that sense, this book does not want to depart from the idea of writing in a rigorous style or from a specific tradition. I believe that any such attempt would miss its mark and, moreover, forget that Serres's own pedagogy of invention is a two-step process: 'To start by being familiar with everything, then to start forgetting everything' (Serres and Latour 1992, 22). Thus to be able to do something genuinely new with Serres, one first has to take up these traditions still at work in Serres's own oeuvre.

In Watkin's book we find a very impressive and illuminating way to get a systematic grip on Serres's philosophy. The central notion to capture Serres's project, according to Watkin, is Serres's notion of 'figures of thought'. Mainly drawn from Serres's book *Le Gaucher boiteux* (2015), Watkin associates it with eight key features of these figures of thought that Serres mobilizes:

> they are operators, they are present in the natural world, they introduce something new into a situation, they arise and are sustained bodily, not just mentally, they are invented and sustained in literature as well as in nature, they are framed as characters with proper names, they synthesise plural features, and they provide a richness of which abstract concepts are only ever a reductive abstraction. (Watkin 2020, 22)

In Watkin's story, the goal of Serres is the development of a 'global intuition', which could be interpreted as coming close to the picture I described earlier: the cultivation and transformation of our set of basic metaphors, myths and schemes by which we interpret the world.

The enemy for Serres is what Watkin calls 'unbillical thinking': 'For Serres, no model is the privileged source of all knowledge . . . but . . . all models are isomorphic with each other and from their complex analogical correspondences there emerges a structure that unites them in, not despite, their differences' (Watkin 2020, 62). Serres will therefore not follow the traditional method of critique (see Chapter 6), but rather 'oppose by generalizing': show that the proposed model to interpret certain phenomena, though not wrong, is not privilege. Countless other models are possible as well.

Though an impressive piece of scholarship, there is a downside to Watkin's approach, shared by many other commentators of Serres. It risks, perhaps, too much to see Serres's work as that of an isolated individual, without any tradition or master. This image is also propagated by Serres himself, for instance in his interview book with Latour: 'What contemporary author have I followed? None, alas' (Serres and Latour 1992, 9). But this picture is a misleading common trope found in French philosophy, especially in its Anglo-American reception. The result can best be described as a kind of *macrophilosophy*: to understand an author, you have to place him within the canon of great authors. Serres is thus constantly compared to Plato, Lucretius or Descartes, as if they were sitting together at the table. To the extent that other French scholars are mentioned in the secondary literature – such as Gilles Deleuze or Jacques Derrida – they are part of a disembodied and timeless Anglo-American canon of continental philosophy in a broad sense.

Against this, this book aims to pose a kind of a *microphilosophy* that would strive to situate Serres's oeuvre in relation to its local neighbours, in our case: French philosophy of science or French historical epistemology. Though open for debate, this tradition spans the whole of twentieth-century French philosophy, and goes back even well into the nineteenth century (see Brenner 2003; Chimisso 2008a; Rheinberger 2010; Bordoni 2017). It is associated with figures such as Léon Brunschvicg, Gaston Bachelard and Georges Canguilhem, or more recently scholars such as Jean-Pierre Dupuy, Bruno Latour, Michel Callon and even the Belgian philosopher Isabelle Stengers.

I believe an examination of how Serres relates to this tradition is warranted, since by not taking it into account any reading of Serres risks to do at least three problematic things. First of all, it can lead to a misunderstanding of what is at stake in some books or claims of Serres. Secondly, it risks ascribing to Serres a set of innovations or discoveries that were not his, but simply part of the tradition in which one could situate him. The result is often that the real innovative elements are overlooked. Finally, by ignoring any connection between Serres and French philosophy of science, scholars from the latter tradition will also not be inclined to engage with Serres's work. Only by showing the connections can a genuine debate be initiated.

Though my claim is not that French philosophy of science is the only way to situate Serres, I nonetheless want to argue that it is a fruitful one. More particularly, I aim to achieve three things. First of all, my goal is to provide the reader with a sketch of a number of debates within French philosophy of science which I believe are illuminating to understand Serres's work. The ambition of the following chapters is therefore not to give a full overview of Michel Serres's philosophy, not even all of his thoughts on science and technology. The aim is instead to offer a guide to help understand a number of recurrent themes and discussions in Serres's books, so that if the reader returns to them, new ways of understanding them become possible.

The second aim is to resituate Serres as part of the tradition of French philosophy of science, replacing the misleading image of a lonely thinker without masters. By highlighting how Serres is connected to this tradition, I also hope to make his work accessible and intriguing to a whole new audience. To do so, I will place Serres between two points in French philosophy of science: the early tradition of 'historical epistemology', embodied by the work of Gaston Bachelard, and more recent Francophone scholarship on science and technology, exemplified by Bruno Latour and Isabelle Stengers. Again, my claim is not that these are the only relevant actors to be discussed (I invoke many others throughout the chapters), but merely that this is an interesting way to situate Serres within this tradition. In a way, the work of Serres can be seen as a potential bridge between two networks of French philosophers of science who otherwise remain unconnected.

These choices, together with the expected limitations of any monograph, also imply that I was forced to leave out a number of authors from my story. For example, I do not extensively discuss Serres's relation to authors such as Henri Bergson (see Delcò 1998, 94–107), André Leroi-Gourhan, Gilles Deleuze (see Herzogenrath 2012) or Gilbert Simondon. I think to do so would moreover require a quite different take on twentieth-century French philosophy, starting rather from philosophy of technology (see Loeve, Guchet and Bensaude-Vincent 2018). I also left out detailed discussions of Serres's views on the nature of the universe or of biological life. Again, this would require a different story situated in philosophy of nature. Serres's work played a crucial role there as well, not only in inspiring thinkers like Ilya Prigogine but also in being criticized for his indeterminist stance, for instance by René Thom, who claimed that Serres's fascination with Lucretius's clinamen (Serres 1977b), 'testifies to an antiscientific attitude par excellence' (Thom 1983, 11). I believe several of the existing monographs on Serres already indirectly take up many of these themes (e.g. Assad 1999; Bühlmann 2020), though not always with the specific French context in mind. My interest is not so much on what Serres had to say about nature or about specific scientific topics, concerning matter, information or life. My interest is in what perspective Serres's work can offer to let us think about scientific *practices* and the place of science and technology in our culture. Obviously, both elements cannot be completely separated from one another. I will therefore touch on Serres's relation to molecular biology, information theory and Serres's connections to the Group of Ten (*Groupe des Dix*), a group of French scientists who were fascinated by all these new developments in physics and biology, related to processes of self-organization.

My final objective is to highlight how, by building on and going beyond French epistemology, Serres's work provides us with a new and fascinating take on science and technology, exemplified in the many fruitful ways in which his work has been taken up by others scholars, such as Latour or Stengers. By this, I want to highlight two things. First of all, that French philosophy of science is more than historical epistemology, even though both are often connected. Secondly, that, though often developed in dialogue with Anglo-American philosophy of science and Science and Technology Studies (STS), one can find within recent French philosophy of science something that is often missing in those other fields: the possibility of a normative stance on science and technology. Throughout the chapters I try to argue that especially one of Serres's central concepts, the *quasi-object*, offers such a new angle. Although we will explore the meaning of the term throughout the chapters, in general it refers to a point of view that abandons the traditional dichotomy between active subjects and passive objects in favor of a more relational approach, where quasi-objects refer to those (human or non-human) entities that constitute and create collectives and social order around them a king, a celebrity, a football, a smartphone or a coronavirus.

By analysing the world through the lens of quasi-objects, Serres opens a perspective where the task of philosophy is to cultivate the right relations, since quasi-objects produce not only order but also violence. The task becomes developing the right *ecology of quasi-objects*: Through which quasi-objects do we want to organize our society? And which procedures do we need for an optimal and fair construction of these quasi-objects? This has become a prominent question given our current ecological crisis that has challenged the way we organize our society, and the technical, economic and political quasi-objects which make this organization possible. The quest for an ecology of quasi-objects thus also becomes a quest for a viable earth, a shift from a parasitic relation with nature to one of symbiosis, to what Serres famously called a 'contract with nature' (Serres 1990).

To enable this ecology of quasi-objects, Serres stresses the importance to cultivate a sensitivity: a sensitivity for the myriad ways in which we are defined by our relations, for the subtle and often invisible forms of violence implied by our societal choices, and for potential alternative quasi-objects to reform our current collective. This 'virtue of sensibility', as David Webb labels it, is often described by Serres in bodily metaphor: it is a training of the body to become more sensitive to the world, learning to be affected by it in new ways:

> To return now to the helmsman, the mountaineer and the musician, their virtuosity consists in sensing more acutely and thereby having more options at their fingertips for how to achieve a more fully developed sense (the course, the sequence of moves or the path, the interpretation). The point is not to develop a more sophisticated interpretation of the data given, but actually to extend the scope of sensibility itself so that more is given before the intellectual exercise of interpretation begins. (Webb 2018, 27)

But, as Webb immediately adds, this 'body' must be seen as extended beyond the flesh: our body can and does also include our technical instruments, through which science and society have learned to become sensitive to the ways in which the world and the critters on it respond to our actions, and we to theirs.

The book aims to explore these themes through two parts. The first part of the book aims to situate Serres in relation to the tradition of French historical epistemology, mainly Gaston Bachelard and his legacy. The second part of the book shifts to Serres's own alternative and its legacy. This part consists of five chapters, each focusing on one dimension of this alternative: materiality, modernity, anthropology, religiosity and ecology. They aim both to highlight Serres's own thought on these subjects and to map how they relate to, have been taken up or transformed by other authors.

The first chapter starts from the common misconception about the relation between Serres and Bachelard, which is often presented as a radical break. By focusing on Serres's early *Hèrmes* series (1969–1980), we will see how Serres portrayed himself initially as a follower of Bachelard, exemplarily shown in his neologism of the 'new new scientific spirit' (*le nouveau nouvel esprit scientifique*), updating Bachelard in the light of more recent scientific developments. This allows for a reinterpretation of the relation between both authors, one where Serres's philosophy can be partly understood as a Bachelardian criticism of Bachelard himself. This Bachelardian criticism consists in what could be called the latter's 'surrationalism': the sciences do not follow the categories imposed by philosophers but are always more flexible and open than these categories allow. Specific critiques by Serres, such as those concerning the novelty of Bachelard's thought, the role of epistemology and finally the political dimension of science will be evaluated through a reappraisal of this Bachelardian move that underlies Serres's criticism.

In the second chapter, I will look at what happened in French epistemology in the 1960s, namely the rise of Althusserianism and its reinterpretation of Bachelard's notion of epistemological rupture as an epistemological break (*coupure épistémologique*). Serres is personally confronted with this movement through Althusser's debate with Jacques Monod. As a result of this historical episode, many recent French philosophers of science, such as Serres, Latour and Stengers, reject Bachelard's idea of an epistemological rupture between imagination and science, but paradoxically praise Bachelard for his insights in the role of technology in the construction of phenomena, mainly through the notion of phenomenotechnique. The chapter ends by problematizing such a strong distinction and the argument will be made that a more interesting reading of both Bachelard and Serres leads to a Janus-headed view on science, where both the element of purification (the epistemological break) and the element of translation (phenomenotechnique) are combined.

Given these similarities, the third chapter aims to explore the central difference between Bachelard and Serres: their respective normative model of how the scientific self should be constituted. Through Foucault's notion of 'techniques of the self', this chapter revisits first Bachelard's and Althusser's ethos of the scientific self, in order to subsequently oppose it to Serres's alternative. Whereas Bachelard proposes a model where the scientific self needs to be made dynamic through a purification from epistemological obstacles, Serres conceptualized the scientific self through the body, which has to be made sensitive by fighting of the numbing noise of the social.

The fourth chapter's focus is on the question of materiality and objects. It starts from a paradoxical reception history of Bruno Latour, who is both accused of paying too little and too much attention to objects. To solve this paradox, François Dagognet's notion of inscription and Michel Serres's notion of translation are invoked. This will lead to a relational ontology, inspired by information theory and molecular biology. While both notions of 'inscription' and 'translation' suggest a philosophy focused on language, we will see how both Dagognet and Serres, and thus also Latour, are at the same time philosophers who call for a return to the object. Therefore, as I will argue, what initially looks like a paradox or a weakness – namely that it is unclear whether Latour speaks about things or about words – is in fact a strength of this alternative perspective, as developed by Dagognet and Serres.

The fifth chapter makes a first step to a new normative approach, mainly by looking at how Serres understands history and modernity. This will be done by comparing both Serres's and Latour's perspective, and confronting them with the diagnosis of postmodernity by Jean-François Lyotard. I will argue that the diagnoses of Serres and Lyotard are in fact very similar, partly because Lyotard was inspired by Serres and Latour. After summarizing Lyotard's diagnosis of postmodernity, the chapter first discusses Serres's own recent attempt to map a Great Story of the Universe. Secondly, it explores Latour's diagnosis that we have never been modern, stressing how it was inspired by Serres's philosophy. Through a reexamination of the work of Lyotard, however, I argue that both these diagnoses are not completely in contradiction with that of Lyotard. Based on this, the second part of the chapter opens up a new dialogue between Lyotard and Serres, centring on the question of an *ecology of quasi-objects*: which relations and quasi-objects one is willing to give up and which not.

Given that quasi-objects are not all equally desirable, the sixth chapter explores which quasi-objects are to be avoided. This will be done through an examination of Serres's anthropology of science. The chapter starts from Serres's notion of 'thanatocracy', which argues that the atomic bomb embodies how we are living in a society ruled by death. But Serres argues that this violence did not start with the atomic bomb, but highlights a deeper anthropology of violence always at work in science. Serres mainly draws inspiration from two thinkers: Georges Dumézil and René Girard. Following Girard's mimetic model of violence, Serres argues that society is always in need of scapegoats to contain this violence in sustainable ways. But, in contrast to Girard, these scapegoats can be non-humans as well. Serres mainly explores these quasi-objects following Dumézil's tripartite hypothesis: quasi-objects in form of a fetish (Jupiter), an army or a weapon (Mars) or commodities (Quirinus). While all three types of quasi-objects put an end to mimetic violence, they also paradoxically create new forms of violence, in the form of sacrifices to maintain this social order. But similar to the Group of Ten (*Groupe des Dix*) and especially Jean-Pierre Dupuy, Serres is also interested in how this logic of sacrificial violence is at work in science and technology, embodied not only by technical objects such as the atomic bomb but also by the academic practice of critique.

The two final chapters explore Serres's alternative to these violent quasi-objects. In the seventh chapter, I look at how Serres connects science and religion through an examination of the metaphor of secularization. Similar to how secularization refers to a decreasing status of religion and God as a transcendent factor in society, the secularization of science refers to an abandonment of Science as something 'sacred' and Nature as transcendent. This chapter explores these secularization metaphors, by arguing for a parallel between how sociologists and philosophers of religion differ and how similar disagreements between sociologists of science and the work of Serres, Latour and Stengers exist. The first part of the chapter explores traditional secularization theories, mainly focusing on Marcel Gauchet and sociologists such as Peter Berger and Thomas Luckmann. It will be shown that the former differs from the latter in an added normative dimension that emphasizes the role of transcendence, even in a secularized world. The second part of the chapter highlights how a similar relation exists between Serres, Latour and Stengers and the sociologists of science. While both agree that a radical transcendence of science from society is untenable, the philosophers nonetheless aim to recuperate some level of transcendence within science by stressing two elements: (a) that scientists deal with quasi-objects, not objects, that always transcend the needs and wants of the scientists; and (b) that these quasi-objects also co-constitute the identities of the scientists and can often not simply be given up. The two points are particularly present in Latour's and Stengers's recent work on Gaia and the Anthropocene, again stressing the question of which relations one is willing to give up. In this they mobilize Serres's notion of religiosity as the opposite of negligence: to be religious is to take the relations valued by others into account.

These questions are also at the centre of the final chapter, which deals with ecology. In the first part, this chapter mobilizes elements from the earlier chapters to present how Serres aims to move away from a parasitic relation with nature to one of symbiosis. In the second part, I argue that the concept of the quasi-object forms the basis of the ecological thought not just of Serres but also of Latour and Stengers: similar to how in science the quasi-objects transcend what we want them to do, in ecological questions we have to negotiate with, rather than dominate, quasi-objects. Based on this a reinterpretation of the famous 'parliament of things' by Latour is proposed as a response to Serres earlier call for updating our social contract to a natural contract. A parliament of things is precisely one that avoids prematurely to purify quasi-objects into objects, but rather installs the required procedures to negotiate and articulate the quasi-objects and the relations they and we require to exist. Technology can play an important role in this, provided we reinterpret its goal as one aimed not (just) at control and purification but also at translation and articulation. This framework offers a particular diagnosis and response to our current ecological crisis and the Anthropocene. This chapter therefore shows how, relying on the insights of the previous chapters, Serres's philosophy of science offers the possibility for a politics of quasi-objects. In that sense Serres's philosophy provides us with a coherent alternative philosophy of our current ecological predicaments.

1

Surrationalism after Bachelard: Michel Serres and *le nouveau nouvel esprit scientifique*

Introduction

The work of Michel Serres is often presented as a radical break with Gaston Bachelard (e.g. Erdur 2018, 339–44). Bensaude-Vincent even describes Serres as 'the one who freed us from the shackles of the Bachelardian vulgate' (Bensaude-Vincent 2010, 43). This antagonistic picture is also endorsed by Serres himself, who in an interview in 1976 described Bachelard's rationalism as 'giant plague in philosophy' which 'has ruined French epistemology' (Serres 1976, 21). In a later interview with Latour, he would similarly state:

> Yes, I wrote my thesis under Bachelard, but l thought privately that the 'new scientific spirit' coming into fashion at that time lagged way behind the sciences. . . . The model it offered of the sciences could not, for me, pass as contemporary. This new spirit seemed to me quite old. And so, this milieu was not mine. (Serres and Latour 1992, 11)

In a similar vein, Bruno Latour has described Serres as the anti-Bachelard (Latour 1991, 93). Within this context the project of Bachelard is described as a naïve belief in the rationality of science or as a misguided project to purify science from all non-scientific elements (see Chapter 2).

This image, however, is too simplistic and in fact makes us unable to really appreciate what we can learn from the work of Bachelard today. As Christiane Frémont (2008, 79) correctly notes, 'from a genuine post-bachelardian one has too hastily made Michel Serres into an anti-bachelardian.' A more interesting picture comes forward if one goes beyond such a simplistic opposition between Bachelard and Serres. By focusing on Serres's interaction with Bachelard in his early work, a core element of the Bachelardian project that is still at work in Serres's philosophy will be highlighted. This does not mean that Bachelard and Serres agree on everything. On the contrary, differences must be recognized, such as Serres's relational ontology (see Chapter 4). While the following chapters will highlight these differences, this chapter deals most of all with the early Serres. This early work shows how Serres's break with Bachelard and French epistemology was produced by an intimate dialogue with this tradition.

In this chapter, the claim is that there is a certain methodological continuity between both authors. It is this methodological aspect of Bachelard that, by still being at work in Serres's philosophy, highlights why it is still meaningful to embed Serres's philosophy in the tradition of French historical epistemology. In the first part, I will discuss the work of Gaston Bachelard, especially his surrationalism and his *philosophie du non*. Secondly, I will use this reading of Bachelard to shed a new light on the specific criticisms Serres has raised against him. The claim is that Serres's criticisms, even when one acknowledges clear discontinuities in content, can be understood as a radicalization of certain methodological elements at work in Bachelard. We will see moreover how these reflections on the relation between Serres and Bachelard are also valuable to the relation between Serres and other representatives of French philosophy of science, ranging from Auguste Comte to Georges Canguilhem. It will therefore also be a step towards the next chapter, where we will focus on Serres's main disagreements with Bachelard's legacy in the 1970s, especially under the influence of Louis Althusser.

The philosophy of science of Gaston Bachelard

There are many different possible attitudes in the philosophy of science. To grasp what is specific to Bachelard's, it is useful to contrast it in a rather schematic way to how philosophy of science is often understood. Philosophy of science seems to be about formulating criteria for how science should behave as a rational process. Bachelard's project is, however, different in several respects. First of all, standard analytic philosophy of science often aims to conceptualize a timeless model of science: that is a model that would work for any specific moment in science whatsoever. Secondly, their aim is to propose norms for how science *should* behave rather than how it factually behaves. In this sense the philosopher has the task of dictating to the scientist how to do science.

Surrationalism and the primacy of science

The programme that we can find in Bachelard is rather different. Its aim is similar in the sense that it wants to understand scientific practices, but it must be seen as part of a bigger project, namely 'writing the history of the mind' (Chimisso 2008a). In fact, in twentieth-century France there was hardly a distinction between philosophy and history of science in the first place. Rather, they have always been intimately related to one another.[1] Central to philosophy of science in France is the idea that to understand the functioning of the human mind, one cannot start from the traditional *a priori* way. One always has to look at the history of the sciences to see the rational movement of thought. At this point, French epistemology comes close to German Neokantians, such as Ernst Cassirer. Though Cassirer was influential in France as well, French philosophy had its own history of Neokantianism, under banners such as spiritualism and reflexive analysis (*analyse réflexive*), connected authors such as Jules Lachelier and Léon Brunschvicg (see Chapter 3).

But an additional reason why authors such as Bachelard follow this approach is linked to the foundational crisis in mathematics and the scientific revolutions in physics at the beginning of the twentieth century (Castelli Gattinara 1998). For many philosophers these crises showed that the traditional assumption of an atemporal foundation for rationality was not so self-evident. How can we still be sure that our beliefs are rational if there can be such historical breaks and revolutions even in mathematics or physics? Projects such as Russell's Logicism or Husserlian phenomenology can be understood as responses to these crises and attempts to find new firm foundations for all rational beliefs.

Following Castelli Gattinara, one could state that in France a different approach was taken. Rather than trying to look for a firm foundation underneath the dust of scientific revolutions, authors such as Bachelard claimed that *rationality was to be found within the revolutionary act itself*. Instead of seeing historicity as a problem for rationality, it was seen as the ground for rationality itself: science was rational not despite its historical shifts, but because of the historical shifts, which were deemed to be rational stages of scientific thinking. The dialectics of the history of science proved the sciences to be rational.

In the case of Bachelard this is argued for in the name of an 'open rationalism' (*rationalisme ouvert*) or *surrationalism* (Bachelard 1934, 175; 1972). For Bachelard rationalism does not imply that one should start from a number of fixed cognitive categories or principles. Any attempt to do the latter he calls a 'closed rationalism', where the forms thought can take are fixed for eternity and in fact limit the way we can think and do science. One could think of a simple Kantian scheme, where rationality is defined by timeless categories of understanding. An open rationalism starts instead from the idea that the act of rationality lies within the overcoming of the categories of thought by creating novel ones, if deemed necessary by the developments of science. Bachelard argues:

> to place reason *inside the crisis*, to prove that the function of reason is *to provoke crises* and that the *polemic reason*, to which Kant had only attributed a subalternate role, cannot leave the *architectonic reason* to its contemplations. We should thus gain access to an open Kantianism, a *functional Kantianism*, a *non-Kantianism*, in the same way as one speaks of a non-Euclidian geometry. (Bachelard 1972, 27–8)

In this sense, similar to the subversive nature of surrealism, *surrationalism* aims to break with conservative tendencies to stick with old categories of thinking. Surrationalism precisely creates the room for scientific practices to redefine our cognitive categories. For Bachelard 'science instructs reason. Reason has to obey science, a more evolved science, an evolving science' (Bachelard 1940, 144). Thus we find in Bachelard the distinctive idea of what I call the *primacy of science over philosophy*: philosophy should not dictate or supervise a normativity of science, but rather learn from the norms internal to the sciences themselves (see Simons 2019).

Against the closed rationalism of the philosophers, Bachelard aims to mobilize an open rationalism. This openness, however, is not found within traditional philosophical activity, but within the scientific practices. As a consequence, there exists a broader tension between philosophy and science within Bachelard's oeuvre. For him, scientists continually revise their own categories, while philosophers tend to be conservative about them (Bachelard 1949, 43). Philosophers wrongly try 'to apply necessarily finalist and closed philosophy to open scientific thought' (Bachelard 1940, 2). For Bachelard, the sciences never follow the clear-cut and given philosophical categories. Rather they create their own novel categories because 'science ordains philosophy by itself' (Bachelard 1940, 22). Or as he states, '[e]very philosophical mind who puts himself to studying science would see how much of contemporary science is philosophical in its core' (Bachelard 1953, 180). Given philosophical categories are in fact never a solution but always a problem. This is related to Bachelard's idea that the formation of the scientific spirit consists in an *epistemological rupture* with everyday experience:

> We believe, in fact, that scientific progress always shows itself in a rupture, in continuous ruptures, between ordinary knowledge and scientific knowledge, as soon as one is faced with an evolved science, a science which, due to these ruptures themselves, carries the mark of modernity. (Bachelard 1953, 207)

We will return to this famous idea of Bachelard in subsequent chapters. Here it suffices to say that this break implies a break not only with psychologically tempting images about the scientific object but also with spontaneous philosophical theories about science. 'The scientific spirit consists precisely in the bracketing of the first philosophy [*la philosophie première*]. Just as the experimental activity, the philosophy linked to the scientific activity must be nuanced and, as a consequence, be mobile' (Bachelard 1951, 17). But traditional philosophy of science does not do this, and therefore 'science does not have the philosophy it deserves' (Bachelard 1953, 20). And this is precisely what Bachelard aims to create in his own work, a surrationalism respecting the openness that is active within scientific practices. In this sense, 'epistemology must thus be as flexible as science' (Bachelard 1949, 10). The implication is that to really grasp what is going on in scientific practices, looking at the history and development of these sciences becomes a necessity.

Scientific practice and *Philosophie du non*

At the same time, however, there is a clear normative idea of scientific progress at work in Bachelard's work. One of his central starting points is that within history of science it is always inevitable to make normative judgements from the present perspective. In this respect, Bachelard contrasts the work of the epistemologist with that of the common historian. The historian looks for facts, and accumulates them in his study without making any normative judgement. This model, however, does not work for the history of science, because 'it does not take into account the fact that every historian of science is necessarily a historiographer of Truth [*de la Vérité*]. The events of science are connected in an ever-increasing truth' (Bachelard 1953, 86). According to Bachelard, such normativity is necessary and meaningful (see Chapter 3). He acknowledges that reading the history of science as a teleological process, where historical episodes must be seen as necessary steps or obstacles with the present as their goal, is problematic. But he distinguishes this from the claim that if one wants to do proper historiography of science it is unavoidable to rewrite the history of science from a contrastive comparison with what is presently seen as scientific and what is not (see Bachelard 1951).

In Bachelard's oeuvre, this normative framework is combined with a clear notion of historical discontinuity about science, embodied by the famous epistemological rupture mentioned earlier. This rupture implies not only a break between spontaneous and scientific concepts but also historical breaks within the sciences themselves. Claiming that the scientific revolutions at the beginning of the twentieth century imply a 'new scientific spirit', Bachelard argues that there exists a radical discontinuity between Newton and Einstein. 'One thus cannot correctly say that the Newtonian world prefigures in its main lines the Einsteinian world' (Bachelard 1934, 46).

At first sight this combination seems problematic. How can contemporary scientific categories be relevant to a past with which scientific practices have broken? Both elements, however, are not irreconcilable for Bachelard, but precisely imply each other. In his book *Philosophie du non* (1940) Bachelard argues that historical progress in science is not made in a continuous manner, but rather through breaks. As stated earlier, it is in these historical shifts that the scientific spirit shows its rationality. But the products of these shifts as well have a specific character, that Bachelard tries to capture through his '*philosophie du non*'.

For Bachelard, the sciences progress through a model of incorporation: there is a radical shift in scientific revolutions, but one where the previous theories are not completely abandoned, but rather reappraised and translated into particular and approximate cases of the new theories. 'The *philosophie du non* will therefore be not an attitude of refusal, but an attitude of conciliation' (Bachelard 1940, 15–16). The example he uses is that of non-Euclidean geometry, which never disproved classical Euclidean geometry, but reappraised it as a specific case of a broader framework. 'The generalization by the no must include what it denies. In fact, the whole breakthrough of scientific thought in the last century has come from such dialectical generalizations resulting in the incorporation of what one denies' (Bachelard 1934, 46). It is in this manner that he speaks of quantum mechanics as a non-Newtonian physics and of his own epistemology as a non-Cartesian epistemology (Bachelard 1934).

The history of science, thus, follows a progressive dialectical movement, which is fundamentally open-ended. The epistemologist must follow and grasp this movement. 'The progress is the dynamics itself of scientific culture, and it is this dynamics that the history of science must describe' (Bachelard 1951, 25). For Bachelard this results in a distinction between lapsed history (*histoire périmée*) and sanctioned history (*histoire sanctionnée*). Since the history of science is not continuous, but cumulative, this implies also the dismissal of certain parts of science. The former is therefore used as a term for these parts of science that, from the contemporary perspective, are excluded as non-science, while the latter refers to those elements that are preserved.

Bachelard thus endorses a specific form of 'presentism' (see Loison 2016). This framework became quite influential in France, and can also be seen at work in the work of authors such as Georges Canguilhem, Serres's supervisor (see Simons 2019). But such a normative presentism is quite problematic for many contemporary historians of science and will also be heavily criticized by Serres. However, I want to argue that in the light of Bachelard's surrationalism, Serres's critique should not be understood as a radical break with Bachelard, but rather as an internal dispute about this methodological principle: Is Bachelard's presentism not in conflict with the very idea of the openness of surrationalism?

Michel Serres's critique of Bachelard

Now that Bachelard's approach has been made clear, we can reexamine the case of Michel Serres and see that the picture of Serres as an anti-Bachelardian has a limited validity only. On the contrary, this reexamination will allow us to stress some interesting continuities at work between the two, hidden beneath the more visible disagreements.

But before going into that, first the question must be raised what precisely is meant with the claim that Bachelard has influenced Serres. This claim might first of all mean that Serres's early philosophy has been greatly shaped by the way philosophy of science was taught in the 1950s, which was indeed strongly defined by the work of Bachelard (see Chapter 2). The typical combination of philosophy and history of science, for example, is found in both authors. But also a clearly constructivist perspective can be found in Bachelard as well as in Serres, focusing not on how the scientist passively studies nature, but rather on how (s)he actively intervenes in nature and 'constructs' the phenomena. Serres might thus be Bachelardian in a rather indirect way, namely through the intellectual climate or through other French historical epistemologists such as Georges Canguilhem, or even through the work of Louis Althusser and Michel Foucault.

But Bachelard himself in particular played a more prominent role, besides being the reference point within discussions about philosophy of science in France. Serres himself was (a) first of all a direct pupil of him, since Serres 'wrote [his] thesis under the direction of Bachelard, on the difference between the Bourbaki algebraic method and that of the classical mathematicians who had gone before' (Serres and Latour 1992, 10). Moreover, (b) Serres discusses Bachelard extensively in his early work, and even in later interviews concerning his philosophical influences. The result is then also that (c) several key concepts were developed in direct dialogue with Bachelard himself. It is for these reasons that it seems warranted to speak of the Bachelardism of Serres.

Besides such simple continuities there are, however, more substantial links in methodology to highlight. For instance, following new developments within physics, Serres develops a non-determinist physics, where the starting point is not order but disorder (see Prigogine and Stengers 1979). This indeterminism, mainly inspired by fluid mechanics and chaos theory, is different from how Bachelard understood physics. However, at the same time this can be seen as a next step in the Bachelardian conception of the development of science, namely that of the *philosophie du non*. Similar to how Euclidean mathematics became a borderline case of non-Euclidean mathematics, Serres develops a perspective in which the old physics is an exceptional case in a broader framework where disorder is the rule. 'Order is not the norm, but is the exception' (Serres 1974a, 49). As Ilya Prigogine and Isabelle Stengers comment on Serres's philosophy: 'There are the rare ones in which determinism exists as a limit-state, costly but conceivable, in which extrapolation is possible between the approximate description of any observer and the infinitely precise one of which Leibniz's God is capable' (Prigogine and Stengers 1982, 150). The *philosophie du non* can thus be found in Serres's reading of contemporary history of science as well. Even though Bachelard and Serres clearly disagree on the content, on a methodological level Serres's distinct view on physics can be seen as loyal to the Bachelardian framework.

Even more telling, however, is that not only the *philosophie du non* but also Bachelard's surrationalism is recognizable in Serres. To dig up this similarity, let's take a closer look at what Serres is criticizing Bachelard for:

> Yes, I wrote my thesis under Bachelard, but I thought privately that the 'new scientific spirit' coming into fashion at that time lagged way behind the sciences. Behind mathematics, because, instead of speaking of algebra, topology, and the theory of sets, it referred to non-Euclidean geometries, not all that new. Likewise, it lagged behind physics, since it never said a word about information theory nor, later, heard the sound of Hiroshima. It also lagged behind logic, and so on. The model it offered of the sciences could not, for me, pass as contemporary. This new spirit seemed to me quite old. And so, this milieu was not mine. (Serres and Latour 1992, 11)

At first sight, it seems that, according to Serres, the whole idea of Bachelard, that in contemporary physics there is a new scientific spirit at work is problematic. But if one looks closely, Serres is not claiming that there have been no shifts at all that call for our philosophical attention. Rather, he seems to claim that Bachelard is lagging behind the newest developments, since in mathematics for instance, 'instead of speaking of algebra, topology, and the theory of sets, it referred to non-Euclidean geometries, not all that new.' The problem is not that Bachelard claimed that there was an epistemological rupture, but that he did not see all the ruptures, or the newest ones. In a way then, we could say, Serres accuses Bachelard of *not being bachelardian enough*, of not being loyal enough to his surrationalism.

Taking this surrationalism of Bachelard as a starting point, the three main criticisms Serres makes of Bachelard can be reexamined. The first criticism is already hinted earlier, namely that Bachelard did not live up to his own standards, because he had not followed the most recent scientific developments that Serres himself witnessed. 'I had the chance to witness, in real time, three or four big scientific revolutions: modern mathematics, biochemistry, information theory and, later on, in Silicon Valley, the digital one' (Serres 2014, 50). Not only does Serres want to open up this epistemological project even further but he also aims to write about a *new new scientific spirit* [*nouveau nouvel esprit scientifique*] (Serres 1969, 1972, 1974a). Serres claims to follow the sciences even more closely, radicalizing a flexibility that even Bachelard lacked. Secondly, these new developments in the sciences result in a shifting role for the philosopher, since the new sciences produce their own internal epistemology. Finally, the problem is not only that the philosophers cannot be epistemologists anymore but also that by trying to be epistemologists, they miss the whole political dimension of science. Epistemologists such as Bachelard failed to hear 'the sound of Hiroshima' and Serres tries to correct this (see Chapter 6). Although these three criticisms seem quite radical, they can be seen in the light of a surrationalist move. In this sense, the claim can be defended that precisely in criticizing the content of Bachelard's project, Serres remains loyal to its methodology.

A new new scientific spirit

As was clear from Serres's description, Bachelard's epistemology is not yet open enough and lags behind contemporary science. In that sense, Serres's early work consists in a correction of this element within the Bachelardian project. Serres considers himself well-placed and even obliged to describe this new new scientific spirit because his own training came not from philosophers, but 'consisted in witnessing – almost participating in – a profound change in this fundamental science' while 'the epistemologists didn't follow' (Serres and Latour 1992, 11).

This new new scientific spirit is mainly inspired by information theory, topology and mathematical structuralism. According to Serres it consists in an ontological shift, resulting in a new ontology that will also prove to be fundamental to his later philosophy (see Chapter 4). Secondly, following the generalization of this ontological claim, Serres also problematizes Bachelard's strong distinction between science and culture, the rational and the imaginary. Inspired by the promises of structuralism, by which Serres was deeply influenced in his early work, he poses the question: a new method is made possible, one that 'excludes nothing; better yet, it attempts to include everything . . . So why would I exclude literature?' (Serres 1991b, 52) Only an *a priori* distinction can prevent us from making this move, a criterion imposed by philosophers from the outside. Again, however, it is by simultaneously dismissing the content of Bachelard's point of view, but radicalizing his methodological principle that we get there.

At several moments, Serres portrays himself as the next step in completing Bachelard's work. In the first book in his *Hèrmes* series, Serres labels Bachelard as the last of the classical projects in literary criticism and the first of the structuralist ones: 'the contemporary idea of critique defines itself relatively easily as a passage to the limit of the Bachelardian incompleteness' (Serres 1969, 21–2). Starting as it does from the opposition between symbolic analyses of images and rationalist studies of truth, Bachelard's double oeuvre is portrayed as the radicalization of this distinction to the point where it implodes: the symbolic analyses of the most abstracts myths, namely the archetypes of nature (fire, water, earth, air). The study of the imaginary then implies a natural history and the study of science a psychoanalysis of images. 'To a false (and original) alchemy correspond true dreams, to a true (and actual) chemistry correspond false images' (Serres 1969, 25n5). Bachelard is the first to combine both projects in one philosophy, but they remain irreconcilable.

Very similarly, in his second *Hèrmes* book, Serres describes a three-stage process: from the subjective-subjective stage of Descartes, through the subjective-objective stage of Bachelard to the objective-objective stage of the new new scientific spirit (Serres 1972, 13). In his recent book on Serres, Watkin (2020, 280–90) gives a good overview of where Serres and Bachelard disagree on this point. But Watkin overlooks the fact that Serres describes this as three *stages*. In this terminology, Serres is referring to the famous law of three stages (or states) of Auguste Comte: the theological, metaphysical and positive stage. Serres wrote multiple texts on Comte (see Serres 1972, 1974a, 1975a, 1977a, 1989). Although often misunderstood, Comte saw the earlier stages as necessary for the birth of positive thinking. By naming them 'stages' [*états*] Comte wanted to stress that they maintained historical relationships to one another. Theology and metaphysics are thus far from worthless, but are desperately needed to achieve positive thinking. Without speculation, no scientific discipline can come off the ground, no matter how erroneous those speculations turn out to be (see Scharff 1995; Gane 2006). In an analogous sense, the position of Bachelard is not one to simply dismiss, but a necessary stage towards the new new scientific spirit.

But the French *état* could also be translated as state, an ambiguity that Serres himself exploits. Comte's law of the three states is, in that way, also linked to the three states of matter. 'The Thirtieth Lesson specifies, about thermology, that there are three states of bodies: gaseous, liquid, solid' (Serres 1977a, 179). According to Serres, in this form the law of the three states does not just concern social history, but natural history as well, for example that of cosmogony: 'in the original furnace the primitive state is gaseous, nebular. Hence the rule for the formation of the world by cooling and condensation, by liquefaction of gases and solidification of liquids' (Serres 1977a, 181). Similarly, Serres will link it to Bergson's critique of science, since the latter claimed that intelligence reduces the fluidity of duration to solid objects. Similarly, there is the famous metaphor of the lump of sugar in Bergson, embodying a confrontation between liquidity and solidity. 'The hidden metaphors of Comte's *Course* are expressed and exploited, become truly heuristic in Bergson. Bachelard will be the last representative of this tradition, sensitive to it in many ways' (Serres 1977a, 182).[2]

It is therefore no accident that Serres focuses on the example of solid objects to map the new new scientific spirit, opposing it to the earlier states of Descartes and Bachelard that reduce objects to liquids or gases.[3] Serres starts from the 'wax' example of Descartes, which is subjective-subjective for him since both the sender (the wax) and the receiver of information (the cogito) possess no fixed and objective information, but rather information that is unreliable and ever changing (as far as Descartes is concerned). For Bachelard, on the other hand, the world is an undetermined realm of complexity and it is the subject that aided by concepts plus a *phenomenotechnique* imposes a certain rigour and thus reliable information on the object. By purifying it from the subjective images related to colour or smell, one makes objectivity possible: 'It will suffice to untangle the naturally confused circumstances in order to truly organize the real' (Bachelard 1934, 172). Hence, we end up in a subjective-objective stage. However, in the light of the above-mentioned ontological shift, the new new scientific spirit goes one step further and is objective-objective, since both the one who studies and that what is being studied can possess, transmit and receive real and reliable information. 'The third stage, one must call objective-objective, since it tends to decipher the language that objects apply to objects by reconstituting, when it is possible, the objective language' (Serres 1972, 94–5). This real information is not limited to the rational side, purified from subjective experiences, but rather consists of both primary and secondary qualities in the object as well as in the subject. One returns to the things themselves; a stage where history and physics, culture and science become one. Real objects, that we encounter in our daily lives, thus regain their recognition (see Chapter 4).

The new new scientific spirit thus results in one unified method to study science and culture rather than in two separated methods. This leads to the typical Serresian readings of authors such as Jules Verne or Émile Zola, in whose work Serres finds the genesis of contemporary scientific theories such as thermodynamics. It is also in this context that one should understand Serres's suggestion that 'the most scientific works, most instructed works of Bachelard, would it not be those concerned with the poetic elements? Would we find here written, through a method of negation and denial, the prophecies of the new new scientific spirit' (Serres 1972, 78)? According to Serres, hidden within the books concerned with the imaginary, Bachelard opens up a perspective that articulates how the material things can impose their information on our minds. Our imagination must therefore not be seen as independent of the world, but rather as part of and in relation to the networks of the world.

However, it is important to note that the claim is not that Zola, for instance, was equally or even better aware of scientific developments than the scientists. Rather, the idea seems to be that one cannot start from a clear distinction between science and culture but should be as flexible as the texts themselves. Scientific theories and ideas can be developed within literary texts as well. Not because there is a hidden layer of scientificity in these texts, but rather because both are part of one network, that is not fundamentally broken in two. Literature can map the structure of certain phenomena according to the same 'logic' as scientific texts, though in a different, but isomorphic register.

Neither does this imply some form of radical relativism, where a scientific practice, theory or text is completely similar to mythical, political or cultural texts. This is not Serres's position. He would allow for differences between a scientific practice and other practices, but never in a radical, ontological and *a priori* way. Claiming that science distinguishes itself by a form of rationalism, even an open one, is already imposing a certain philosophical category on science, namely that of rationality. Instead, Serres wants to separate the question of the *rationality* of the sciences from the underlying question of what one could, following Stengers (1993), call the *singularity* of the sciences: What is the specificity of scientific practices that distinguishes them from other types of practices? Referring to the rationality of these practices is a possible answer to this question, but one that cannot be given *a priori*. That would imply a form of violence, namely that of purifying our idea of science from certain elements without due process (see Chapters 2, 3 and 6). Instead, one should follow the sciences, even through myths and literary texts, rather than delineating beforehand what the limits of the sciences are. In this sense there is a clear discontinuity with Bachelard, namely by abandoning the notion of rationality. Of the surrationalism, it is the *sur-*, the open movement that remains at work in Serres, while he abandons the *rationalism*.

It is in this light that one has to understand the radical shift in how to read the history of science. Serres disagrees with the picture sketched earlier by Bachelard, that there are clear epistemological ruptures that break with the imaginary part of our thinking and that there is a meaningful distinction to be made between lapsed history and sanctioned history. Such an *a priori* imposing a philosophical dichotomy precludes us from following the sciences in their totality. It prevents us, for example, to see 'contemporary' science at work in the work of authors such as Lucretius. In his book on Lucretius, Serres tries to show exactly how his work, often seen as part of lapsed history, has reemerged as relevant for contemporary fluid mechanics. 'Scientific modernity does not enter history by a fault or a break, but by the revival of a philosophy of nature that has been spreading ever since Antiquity. The so-called break is an artefact of the university' (Serres 1977b, 41).

The new new scientific spirit thus results in an alternative epistemology of science, which could be called the *model of proliferation*. The central idea is that the objectivity of science is *strengthened* if it is linked to more elements and connections. Elements of literature and imagination can thus play a positive role in the production of knowledge. This is opposed to the *model of purification,* ascribed to Bachelard, where science becomes more objective if it is purified from imagination and epistemological obstacles (see Chapter 2).

But there is an ambiguity here, in the case of Serres. He characterizes science which follows the model of purification as repressive with respect to an original multiplicity. 'Let this scientific knowledge get rid of its arrogance, its masterly, its ecclesial dispensation, let it abandon its martial aggressiveness, the hateful pretension of always being right, so that it speaks truth, that it descends, pacified, towards common knowledge' (Serres 1982a, 20). At the same time, Serres claims that this multiplicity can be adequately articulated in his own model of proliferation. In his book on Lucretius, for instance, Serres claims that modern physics is closed off in laboratories, while the fluid mechanics found in Lucretius also works outside the lab, capable to grasp the multiplicity of the world itself (Serres 1977b, 68). His book has indeed been read as 'a story in which physics neither represses (through experimentation) nor manipulates nature' (Harari and Bell 1982, xvi). Serres, thus, paradoxically, believes to have found a model that is no model, a model that contrary to all other models does not reduce, repress and push into categories the original multiplicity found in nature.

But such a belief in a 'model without a model' seems unwarranted. A more plausible view is to acknowledge that all models imply some form of violence towards their objects, but not all in the same way. The model of purification, thus, can be criticized for implying a repression of the purified phenomena, but the model of proliferation is itself not free from this violence. It (hopefully) implies *less* violence or violence of a more acceptable sort. To evaluate this, it is probably necessary to relate the issue to the specific phenomena one is talking about. Serres speaks about whirlpools or climate models. For these phenomena a model of purification seems problematic. But perhaps Bachelard's model of purification can still be applied to a number of cases, such as quantum mechanics or the theory of relativity. Again, one might mobilize Bachelard's *philosophie du non*: we can interpret the model of proliferation not as complete dismissal of that of purification, but instead as a model of non-purification (in the Bachelardian sense), where classical purification remains a limiting case in the broader framework.

One possible response to this criticism of Serres is found in Watkin (2020). As Watkin argues in his book, the typical strategy of Serres is to *oppose by generalization*. Serres expresses his disagreement with specific models, not by criticizing them, but by showing that they are merely one instance of an infinite set of possibilities. For instance, in his dissertation on Leibniz, Serres does not criticize Russell's (logical) model of Leibniz' work as wrong, but diagnoses it as merely one way of approaching Leibniz' oeuvre. What is false is not so much the specific model, but its pretence that it is the only way to interpret Leibniz. By proliferating alternative ways of organizing parts of the world, Serres destabilizes the existing ones. Watkin captures the power of such a conceptual move very well through the following example:

> In the folktale Young Tom catches a leprechaun, who is thereby obliged to reveal the tree in the forest under which a great hoard of gold is buried. Tom marks the spot by tying a ribbon round the tree, and the leprechaun promises not to remove the ribbon before Tom arrives back with his spade and wagon to take away the gold. When Tom returns, he finds that every tree in the forest now has an identical ribbon, and the leprechaun has vanished. In a similar way, opposing

by multiplying subverts without polemics, without negation, without dialectics. (Watkins 2020, 90)

Serres's alternative is then not so much the absence of a model, but the juxtaposition of all models next to one another, while refusing to privilege one of them.

A new image of the philosopher

The new new scientific spirit also forces Serres to accept a new role for the philosopher. First of all, Serres introduces a new term to describe the philosopher. Again inspired by Auguste Comte, the model that Serres prefers is that of the *encyclopedist*, collecting different sciences and types of knowledge, without reducing them or forcing them in a strict hierarchy. Rather they are situated next to one another, with the everlasting possibility of cross-references (Serres 1969, 70). 'Science is, on and for itself, a collection of dictionaries: The Encyclopedia' (Serres 1972, 38). In this sense, we should correct our claim that Serres follows the model of the *philosophie du non*. The model of the encyclopedia is a different, yet radicalized version of the *philosophie du non*. First of all, in the sense that, although regional rationalisms are clearly also present in Bachelard (1949), in the case of Serres these regional criteria for knowledge and truth are also internally developed by the sciences rather than conceptualized by epistemology (see the following).

Secondly, the model of the encyclopedia starts from a different image than that of conciliation and dialectics. 'The new spirit focused itself in a philosophy of no; the new new spirit develops itself in a philosophy of transport: intersection, intervention, interception' (Serres 1972, 10). Instead of clearly demarcated disciplines, the new new scientific spirit follows a model of translation and exchange: 'The singular disciplines become exchangers: of concepts, of methods, of models' (Serres 1972, 65). Again information theory is the paradigm here, for instance in the case of molecular biology. What molecular biologists show is that in genetics one should not search for the 'noumena' underlying our biological beings, the invisible behind the visible, but rather the universality of the genetic code (Serres 1974a, 21). Notions from information theory are thus translated, and at the same time transformed, when applied to a different scientific region, in this case biology.

The encyclopedist, moreover, is not just the Bachelardian epistemologist updated by insights from the contemporary sciences. These newest developments within the sciences have also resulted in a qualitative shift in the sciences themselves, problematizing this traditional image of the epistemologist. Serres radicalizes Bachelard's claim that the sciences themselves produce philosophical categories, by claiming that they now also produce their own epistemology. Serres makes this claim first of all for mathematics:

> At all moments of grand systematic reconstruction, the mathematicians become the epistemologists of their own knowledge. This transformation is a mutation that is carried out from the inside out. Everything happens as if, at the moment of promoting itself into a new system, mathematics suddenly needed to import the totality of epistemological questions. (Serres 1969, 68)

We will see in the next chapters that Serres makes similar remarks about contemporary biology and physics, for instance about the work of Léon Brillouin (1956): 'The philosophers do not have to search nor write a manual where one would find the epistemology of the experimental knowledge. It is there' (Serres 1974a, 45). This has crucial consequences for the task of the philosopher. The idea is that, even if the traditional epistemological project succeeds and one is able to describe the scientific practice in a genuine way, one would only be repeating the sciences themselves. If so, in what sense, then, does the 'philosopher's work differ from that of a journalistic chronicler, who announces and comments on the news' (Serres and Latour 1992, 15)?

Or more precisely, the philosopher is confronted with a fundamental choice, which Serres at one moment compares to literary criticism. The literary critic can choose either to describe the text as loyally as possible, but he will end up in a philological exercise that will not really add anything significant; or else he tries to be more speculative, but he loses himself in an uncertain art of describing, linked to a certain normative framework. So either epistemologists merely repeat the sciences or they become speculative, in that case implying a tension with the original idea of the primacy of science (Serres 1969, 62). Here already it is clear that it is by pushing the traditional Bachelardian project to its boundaries that Serres arrives at one of his fundamental differences with Bachelard: if one would really take the surrationalism of Bachelard seriously, then one can no longer write the books Bachelard wrote.

From an epistemology to a political philosophy

However, Serres does not choose to give up all speculative ambition, but rather the opposite. One can never escape the speculative element, and it is therefore necessary to be explicit about it. For Serres, this means opening up for the political side of science as well. An encyclopedist would only add something to the internal epistemology of the sciences if he or she would speak of more than mere epistemology. 'To speak the truth, my interest in the relations between science and society marked at the same time my difficulties with philosophy, and, most of all, with Canguilhem and Bachelard. They were out of their time. How could one teach epistemology of physics while omitting deontology' (Serres 2014, 160–1)? Serres thus goes further than a mere epistemology in the traditional sense, switching to a political project, which aims to correct Bachelard's project by articulating the political violence of the sciences as well (see Chapter 6). But even this break can be read as playing out Bachelard's own cards against himself: if you really want to pursue an open philosophy, then you must also make room for the political and violent dimensions of science.

But it is important to fully understand the multiple reasons for this shift in Serres's work. In the remainder of the book, we will explore a number of societal and scientific shifts that Serres often invokes to justify such a shift. But it would be too easy to merely reduce this shift to an externally caused event, which Serres noticed, but others allegedly failed to see. There were additional reasons that made Serres look for other angles and issues. A first set of reasons is biographical. We already mentioned that Canguilhem was one of the supervisors of Serres's dissertation. Canguilhem was typically seen as the successor of Bachelard, taking over his position at the Sorbonne, for example. Serres would defend his dissertation on Leibniz in May 1968 (see Serres 1968a), in front of an impressive jury: 'Georges Canguilhem, specialist in scientific epistemology, and Jean Hyppolite, specialist in the history of modern philosophy, who were both my supervisors, but also Suzanne Bachelard, Gaston's daughter, and Yvon Belaval, who was a specialist in Leibniz' (Serres 2014, 48–9). Serres and Canguilhem seemed, at least initially, close to one another. In his dissertation, for example, Serres thanked Canguilhem for filling in the empty spot of his own father, who died early (Serres 2014, 49). His secondary thesis, under the supervision of Canguilhem, *Essai sur le concept épistémologique d'interférence*, clearly saw itself as a continuation of the epistemological tradition and would form the basis of one of his *Hèrmes* books (Serres 1974a).

However, something happened that broke that relationship, somewhere around the day of the defense. Though Serres regularly mentioned this in interviews, he refused to go into detail: 'Let's say there was a tragic moment in my personal and academic history that I don't like to talk about. Until then, Canguilhem had taken me under his wing. I was his favorite student of sorts.

He got mad at me that day' (Serres 2014, 49). It is hard to reconstruct what happened, but it seemed to be related to Canguilhem finding Serres's work somewhat arrogant, while not sufficiently recognizing his debt to the tradition of French epistemology, especially to Bachelard. In *Idéologie et rationalité dans l'histoire des sciences de la vie* (1977), Canguilhem mentions Serres as one of two recent challenges to historical epistemology. The first challenge is Dominique Lecourt (1969) who argues that 'Bachelard failed to free himself from the toils of idealist philosophy, when he should have seen that his conclusions were actually consonant with the doctrine known as dialectical materialism' (Canguilhem 1977, 17). This will lead to the Althusserian reading of historical epistemology (see Chapter 2). 'Another young epistemologist, Michel Serres, raises a different objection. The history of science, he says, does not exist' (Canguilhem 1977, 18). Canguilhem reads an early text of Serres (Serres 1974c) as arguing that before we can have a general history of science – in contrast to local histories of specific disciplines – we first need a 'critical history of classifications', inspired by Serres's own work on Comte (see later). Canguilhem, however, is sceptical and wonders: Is Serres not reinventing the wheel? Referring to Bachelard's *Rationalisme appliqué* (1949), especially the chapter on regional rationalism, Canguilhem argues 'that Bachelardian epistemology confronted this problem well before anyone had thought of accusing historians of ignoring it' (Canguilhem 1977, 18).

A second element, perhaps partly caused by the first, was that Serres felt exiled, since he was unable to get a position in philosophy, instead ending up in the department of history of science:

> I got a teaching position the following year [1969], but in the history of science. I was told it would be temporary, but in reality I found myself banned from philosophy at university. I had to teach outside of my profession. I used to have five hundred people in my philosophy class, and at one time I only had a handful from history of science . . . Fortunately, I was able to find refuge in the United States, at the universities of Baltimore, Buffalo, New York, and finally from the 1980s, Stanford. (Serres 2014, 51)

When recounting a number of other events, Serres reiterated this idea that he felt alienated from his fellow French philosophers. There is the story that Serres recounts of a dinner with a number of other philosophers, including Martial Gueroult, Georges Canguilhem, Louis Althusser and Michel Foucault. During the evening Foucault proposed a game of 'telling the truth', concerning the question: What would you have wanted to be, if you had not been a philosopher?

After Canguilhem insisted the answer should be submitted by secret ballot, Serres reported:

> Each of us took his sheet of paper, and I wrote 'Even still, a philosopher'. When the seven votes were opened, all of them except mine said 'Minister' . . . I found that tragic. Pathetic. These guys confessed that finally they had only one desire: power. . . . There was some discussion and it transpired that the most interesting job was Minister of the Interior. Not Foreign Affairs, not Justice, not Education. Minister of the Interior. Unanimously. (Serres 2014, 52)

Similarly, he reported how 'Claude Lévi-Strauss and François Jacob wanted me to join the Collège de France because they liked what I was doing', an effort which failed because 'the philosophers opposed it' (Serres 2014, 54) Finally, Serres recounts a final meeting in 1995 with Canguilhem, just a few months before the latter died: 'He just asked me at the end if I had had a lot of PhD students in my life. I replied: You know, sir, that I never had one since I was not in my discipline. I was teary-eyed . . . and so was he' (Serres 2014, 51).

But more than mere biographical facts must be invoked to understand Serres's divergent position. There were also intellectual shifts happening in French philosophy in the 1960s and 1970s Serres could not identify with. For instance, it would be incorrect to say that a political project is completely absent in the case of Bachelard, but present in that of Serres. In the next two chapters, we will see that Serres argues that in Bachelard's oeuvre there was always something like a political project. This project is exactly his presentism, the fact of looking for epistemological obstacles and epistemological breaks. For Serres, the model of purification is not only crippled epistemology but a crypto-normative project where true science is seen as that which purifies itself from all obstacles, from imagination, from myth. Bachelard's *La formation de l'esprit scientifique* (1938a) is a political project of how the scientific city should be arranged, namely one with strong walls against imagination, in favour of a spiritual purification of the scientist. Against this Serres states that 'a totally purified reason is a myth' (Serres 1972, 210) and that in fact 'there is no purer myth than the idea of a science purified from all myth' (Serres 1974a, 259).

Serres would apply the same analysis to other French philosophies of science, such as the project of Auguste Comte. For instance, one of his *Hèrmes* essays opens with the claim: 'We lack an Auguste Comte' (Serres 1974a, 159). Such an Auguste Comte is needed due to the problem of the classification of the contemporary encyclopedia. Since we have no new Comte, we still tend to fall back on his old scheme to classify current science: 'without citing him, everyone will repeat the old . . . . The old institutional walls preserve it and endlessly make its ghost reappear' (Serres 1974a, 159). We, however, lack a new framing of our sciences that does justice to its current status, in the form of the new encyclopedia described earlier.

However, this does not mean that there is an objective way to classify the sciences. Instead, the latter is a political project. The same questions that came up in relation to Bachelard reappear here: If Comte introduces a classification of all the sciences, what was the status of that classification? If the classification itself would be scientific this would entail a circle, since this new science should be situated in the classification itself. Similar to the Platonic problem of the third man, this entails 'the argument of the third science' (Serres 1972, 19). Serres indeed argues that Comte's classification seems far from scientific. He even goes so far to state that Comte's 'encyclopedia of exact sciences was dead from the moment it was born' (Serres 1974a, 159). According to Serres, Comte precisely excluded that which turned out to be crucial for the new new scientific spirit: logic and huge parts of mathematics, for being too abstract and static; a cosmology concerning everything outside of our solar system; statistics and speculations about all phenomena not visible to the eye; and so on. It is clear that on all these topics, we have done precisely the opposite. In line with Bachelard's *philosophie du non*, we should therefore be 'non-positivists':

> On the balance sheet, the following are non-positivist (and it is Auguste Comte who said 'non'): our logic, for existing, our regional scientific languages, for developing on the fringes of everyday language, our mathematics, for being abstract, algebraic, formalized, axiomatic; our astronomy, for giving ourselves the universe as an object, and as a method all the physical and chemical disciplines, in an original intersection; our thermodynamics for using random functions; our optics and our electricity, for postulating photons and electrons, and so on. In conclusion, all of our science is non-positivist, it is formed from the prohibitions of the *Course*. (Serres 1974a, 167–8)

If Comte's scheme is thus not scientific, the alternative is to argue that the classification originates from somewhere else, being a product of ideology or culture. Serres even links it to traditional, mythical or even 'primitive' acts of classification, making it the object of the ethnologists more than the positivists. 'In a new twist, it is no longer for the epistemologist to examine his tableau, to learn anything about science, but for the ethnologist, who will learn about what our culture can be' (Serres 1972, 20). But the first paradox reappears here, since ethnologists themselves can only classify based on an idea of scientificity, perhaps even falling back in the end on the scheme of Auguste Comte. 'I do not know whether it is appropriate to place Lévi-Strauss behind Comte or Comte behind Lévi-Strauss' (Serres 1974a, 169).

In fact, Serres anticipates a number of recent scholars (Scharff 1995; Gane 2006) who have argued that Comte's classifying project indeed mainly had political ambitions. Similar to Bachelard, Comte's project is thus not a break with culture, ideology or myth, but instead a continuation of it. A genuine classification of the science cannot be an external reflection on it, such as the epistemologist wants. 'Whoever speaks well of order is situated within that order, or his language is poorly formed; the science of science is either one of the sciences – a partitive genitive – or it is not a science at all, but instead a political project' (Serres 1972, 20). Implicitly, therefore, the epistemological project carries a social and political philosophy, which must be unearthed. As we will see in the next chapter, this model is not restricted to Bachelard or Comte but also at work in those who were inspired by them, such as Georges Canguilhem, but especially Louis Althusser and his pupils. The latter politicized Bachelard's concept of the epistemological break in the name of Marxism and its ideal of a science purified from all ideology.

Conclusion

This chapter has tried to show how the work of Michel Serres must be seen not as an 'epistemological break' with Bachelard but rather as a specific and critical continuation of certain of its methodological elements. Once again, the claim is not that there are no serious disagreements between both authors, only that Serres's work can be read in line with some of Bachelardian notions such as *philosophie du non* and surrationalism. To do so, one has to look at Serres's early work, where he tries to develop a *new new scientific spirit*, updating Bachelard's *new scientific spirit*. But besides mere epistemological corrections, this also implied a shift in the role of philosophers, who have to open themselves to political issues following from science. These shifts have been quite radical, so much so that Serres's recent work is often quite far removed from what one should associate with Bachelard and the tradition of French historical epistemology. But the claim is that it was precisely due to the insights Serres found in the new new scientific spirit, which opened up room for imagination and culture in science, that he saw the need for a political dimension of philosophy of science.

2

Michel Serres and the epistemological break

Introduction

In the first chapter we saw how the traditional picture of a break between Bachelard and Serres had to be corrected. But I do not wish to deny that both authors disagreed on fundamental points. In this and the next chapter we will therefore sketch an alternative story of how Serres slowly distanced himself from this earlier tradition of French epistemology. We will do this in two ways. In this chapter, the focus will be on the broader intellectual context of French philosophy. This will form the basis to explore, in the next chapter, how Serres's project differs from Bachelard's and how that difference is mainly a normative one.

Our starting point will be the ambiguous reception history of Bachelardian epistemology. One of the most productive notions introduced by Gaston Bachelard within the philosophy of science and technology is that of *phenomenotechnique*. With this concept, Bachelard intended to highlight the constructive aspect of contemporary sciences. 'Science does not correspond to a world to be described. It corresponds to a world to be constructed' (Bachelard 1951, 46). An astonishing fact, however, is that several recent scholars who endorse this constructivist claim are at the same time profoundly critical of the work of Bachelard. This is especially clear in the work of Michel Serres, but also of that of Bruno Latour and Isabelle Stengers. For them, science studies and historical epistemology are radically opposed in their programmes (see Tiles 2011). They are especially critical of another aspect of Bachelard's work, associated with his concept of the *epistemological break*. The result is a philosophical puzzle: How can one be a fan of phenomenotechnique while distancing oneself completely from Bachelard and historical epistemology?

The answer to this puzzle depends on how phenomenotechnique and the epistemological break relate to one another. In fact, there are several possible answers, each with their specific consequences. The first fundamental question is whether both concepts are intrinsically linked within the work of Bachelard. If they are in fact distinct, then obviously one can maintain this double stance. But, if they are linked, a second question pops up: Does their link prove to be the fiat of the notion of phenomenotechnique, since the baggage of the break cannot be maintained, or rather a reappraisal of the notion of phenomenotechnique itself?

It is this last claim that I wish to defend. Although superficially Serres and others seem to endorse the notion of phenomenotechnique while dismissing that of the epistemological break, their position is more complex and interesting. In order to articulate this more complex position, specific episodes of the history of Bachelardism must be taken into account. First, it will be explored how the notion of the epistemological break has mainly been taken up from the 1960s in the Althusserian school. One finds the term in the work not only of Althusser but also in that of his followers, such as Étienne Balibar or Dominique Lecourt. More specifically, we will look at how Serres experienced that intellectual development, embodied by Althusser's criticism of the work of Jacques Monod. On the other hand, the notion of phenomenotechnique has mainly gained ground from the 1980s, in discussions within fields such as STS or in the work of authors such as Bruno Latour, Ian Hacking or Hans-Jörg Rheinberger, while they either criticize or ignore the notion of an epistemological break.

In the second part of this chapter, however, it will be argued that such a split-up stance is hard to maintain, both in the work of Bachelard and in the work of more recent Francophone authors, including Serres. In their work both concepts are interconnected and presuppose one another. The conclusion, however, is not that it would therefore be impossible to mobilize one concept without the other. One could of course stick to just one concept and ignore the other. The real conclusion is that it is possible to mobilize both concepts, and moreover that such an endeavour leads to a more complete and interesting picture of science. One can see them as two faces of phenomenotechnique, namely that of *technologies of control* and *technologies of negotiation* (see Chapter 8). Science, in this sense, is Janus-headed.

The epistemological break and the Althusserians

Let me first start with the notion of the epistemological break, which played a crucial role in the reception of Bachelard's work in the 1960s and 1970s. Especially in France, Bachelard's work had mainly been taken up at that time by Louis Althusser and his students. They mainly focused on his philosophy of science, resulting in a neglect of scholarship on Bachelard's books on the imaginary in France (see Wunenburger 2012). But even if one sticks to his epistemological work, one notices another one-sidedness at work, namely that of an exclusive focus on the epistemological break at the expense of phenomenotechnique. As we will see, this specific history is one of the main causes for this strange contemporary attitude towards Bachelard, and for Serres's relation to the tradition of French historical epistemology.

To understand how Althusserians mobilized Bachelard, it is important to have an idea of what Bachelard himself had in mind with the notion of 'epistemological rupture' (Bachelard 1949, 104). A first thing to note is that Bachelard uses this concept in a very specific setting, both spatially and temporarily. An epistemological rupture always refers to specific periods of specific sciences, rather than to science in general. It is for instance claimed for the case of chemistry in the nineteenth and twentieth centuries (Bachelard 1953, 6–7). Secondly, such epistemological ruptures are claimed in two senses, what could be called a *vertical* and a *horizontal* rupture. Either it refers to the vertical rupture between ordinary and scientific experience. 'Between ordinary knowledge and scientific knowledge there seems to us such a strict rupture that those two types of knowledge could not have the same philosophy' (Bachelard 1953, 224). Horizontally, the claim is also made that there are clear discontinuist shifts within the history of science. The most famous one is the 'new scientific spirit' (Bachelard 1934) starting in 1905 with the publications of Albert Einstein (Bachelard 1938a, 9). Such a horizontal rupture refers mainly to historical shifts within sciences that make progress not by accumulation but by rational reorganization, for instance physics after the theory of relativity and quantum mechanics.

From the 1960s on, however, Althusser and his students started to reconceptualize Bachelard's notion of epistemological rupture (*rupture épistémologque*) under the banner of an 'epistemological break' (*coupure épistémologique*) in order to grasp the scientificity of Marxism (Althusser 1965; Lecourt 1975; Balibar 1978). According to Althusser, there is a clear epistemological break in the work of Marx between his early, ideological work and his mature scientific work. By this break Marx discovered the 'continent of history', similar to how the Greeks founded the 'continent of mathematics' and Galileo the 'continent of physics' (Althusser 1965). To make this claim, he mobilizes historical epistemology in a specific way:

> To understand Marx, we must treat him as a scientist among others, and apply to his scientific oeuvre the same epistemological and historical concepts that we apply to others.... Marx appeared thus as the founder of a science, comparable to Galileo and Lavoisier. Moreover, to understand the relation that the work of

> Marx maintains with the work of his predecessors, to understand the nature of the break [*coupure*] or the mutation that distinguishes him from them, we must interrogate the work of the other founders, who as well have broken with their predecessors. The intelligence of Marx, the mechanism of his discovery, the nature of the epistemological break that initiates his scientific foundation, brings us to the concepts of a general theory of the history of the sciences, capable to think the essence of these theoretical events. That such a general theory does not yet exist but as a project, where it partly has taken shape, is one thing; that such a theory is absolutely indispensable to the study of Marx, is another. (Althusser and Balibar 1965, 16)

Althusser and his students thus reinterpret this notion of the epistemological break in a bold way, separated from Bachelard's specific historical cases. Althusser's aim was to establish 'historical epistemology and the history of science as a regional field within the general science of history [i.e. historical materialism]' (Resch 1992, 181). This is, for example, clear in the 1967 lectures Althusser and his students gave for scientists, which consisted in the exposition of a *general* model of the relations between philosophy and science (see Macherey 2009). The task of philosophy, for Althusser, is to make distinctions between the ideological and scientific elements in specific practices and discourse. It resulted in an abstract view on scientificity, used and abused to criticize any opponent as 'ideological' (Althusser 1974a).

A comparable message is found in the work of his students. Michel Pêcheux and Michel Fichant, for instance, defend what they call the 'discontinuous position', which according to them 'rejects the notion of "knowledge" as continuous development, from "ordinary knowledge" to "scientific knowledge", from the dawn of science to modern science' (Pêcheux and Fichant 1969, 9). The result is a radical *a priori* dismissal of all other philosophies of science as idealist ideologies, for instance the 'continuist' positions of Pierre Duhem or Léon Brunschvicg (Pêcheux and Fichant 1969, 74). Similarly, the distinction between 'sanctioned' and 'lapsed history' of science that Bachelard introduced at one point (see Bachelard 1951) becomes for them a general criterion to judge any history of science whatsoever.

And although such a project can be praiseworthy, it has had clear historical effects. The major result is that the epistemological break is transformed into an abstract general theory of the history of science (Pêcheux and Fichant 1969, 101). There is no room left for the specifics of the sciences and no word on the role of technological instruments is uttered. This reading has coloured the reception of Bachelard in France, but even more that of the Anglo-American world. One of the first introductions to Bachelard translated into English was a book by Dominique Lecourt, who reinterpreted the work of Bachelard from an Althusserian perspective (Lecourt 1975). The result is that other more recent literature on Althusser, when, discussing Bachelard, similarly only speaks about the epistemological break (e.g. Resch 1992, 178–81). This has also been noted by commentators in France, such as Canguilhem: 'Italian, Spanish, German, and even English readers have come to know Bachelard's work not firsthand but through translations of critical commentaries, particularly that of Dominique Lecourt' (Canguilhem 1977, 11).

My claim is that this Althusserian Bachelardism is responsible for the typical image of Bachelard's work that is being criticized in the work of Serres, and subsequently also by Latour and Stengers. Let us call this the *model of purification*. According to this model, science consists in a purification of the scientific self from all 'epistemological obstacles' (Bachelard 1938a). Such obstacles are to be understood as images arising out of human imagination or, in the Althusserian reading, reflections of the dominant ideology. Within this picture, science implies a different way of thinking, free from ideology or imagination, free from opinion and myth.

In their criticisms, these authors link this model of science with Bachelard's *La formation de l'esprit scientifique* (1938a), which Serres sees as embodying a specific normative model of science, aimed at the purification of the soul of the scientists from all imaginary traces (see Chapter 3). In the same regard one must see the criticisms of Serres that 'there is no purer myth than the idea of a science purified of all myth' (Serres 1974a, 259). Serres would reiterate his discontent with such a purification, linked to Bachelard, in later texts:

> our generation had learned, once again, at the school of Gaston Bachelard, that we had to confine the elements, air or fire, earth and water, the components of the climate, to the dreams of a vain and lazy poetry: on one side, canonized knowledge, epistemology, awake reason at work, on the other, the imagination, tolerated provided that it remain on the exterior, on the side of sleep and the humanities, which are judged to be dreamlike. (Serres 1994, 92)

Latour, very similarly, states in an interview, 'when I read Bachelard, *La formation de l'esprit scientifique*, I felt vaguely that everything inside was false, anthropologically false, that unreason could not precede reason in that manner' (Latour 2003, 66). Finally, Isabelle Stengers ascribes to Bachelard the 'disqualification of the "non-science" . . . associated with the notion of "opinion", which "thinks badly", "does not think", "translates the needs of understanding". Thus science is always constituted "against" the obstacle constituted by opinion, an obstacle Bachelard defined as a quasi-anthropological given' (Stengers 1993, 26).

My claim is that it is a misunderstanding to see these criticisms as ahistorically aiming at the work of Bachelard. One has to understand the historical situatedness of these remarks, namely as a response to how Bachelard was being used by Althusserian inspired epistemologists in France. That Serres and others, for example, systematically use the notion of epistemological break (*coupure*) and not epistemological rupture already indicates that Althusserianism, and not Bachelard, is (or should be) the target. Let us first focus on how Serres specifically engaged with this Marxist development.

Serres, Monod and Althusser

The relation between Michel Serres and Marxism has not really been studied so far. Only Canguilhem seems to make the suggestion that Serres might be inspired by Marxism, mainly in his early use of the concept of ideology. 'The use of these terms would seem to imply a Marxist point of view, but the context is unclear' (Canguilhem 1977, 18). And though Canguilhem also refers to a passage from Serres's *Esthétiques sur Carpaccio* (1975b, 86–8), Serres's relation to Marxism is not further explored. However, it seems crucial, especially to understand Serres's negative attitude towards many French epistemologists.

Despite these remarks by Canguilhem, Serres mainly seems to be dismissive of Marxist philosophy. As stated in a later interview, he saw more value in the utopian socialists: 'I have great esteem for the so-called 'utopian' socialists who have left us with things that make life easy: mutual funds, nurseries, social security, while so-called 'scientific' socialism has murdered millions' (Serres 2014, 259). This attitude mainly seems to be a product of the intellectual context in Paris, where Serres studied. At the *École Normale Superieure* there seemed to be a general 'political commitment, rather Marxist' (Serres 2014, 37) from which Serres distances himself, claiming that 'this milieu was not mine' (Serres and Latour 1992, 11).

It was Althusser who was setting the tone of the debate around science in Paris when Serres was studying there. Serres was mainly unconformable with Althusser's criticisms of those new sciences, ranging from quantum mechanics to molecular biology, that Serres himself saw as the exciting new developments. He would later report how, when he invited Louis de Broglie to give a lecture to the philosophers in 1954,

Broglie was threatened when he came to lecture at the *École Normale*. I saw him leave the place under the protection of two or three persons because the Marxists were attacking him. These Marxists did not understand a single word of physics. They didn't know what falling bodies meant. And they claimed that Schrödinger was defending a 'bourgeois' science! (Serres 2014, 192)

More particularly, in the 1950s and 1960s, there was the international Lysenko affair. In general it refers to how due to a political campaign Trofim Lysenko gained the favour of Joseph Stalin, and as a consequence Soviet genetics was suppressed in favour of a form of Lamarckism. Genetics was seen as a 'bourgeois' science, in opposition to Lysenkoism which was a 'proletarian science'. Opponents were dismissed, imprisoned or even executed, while Lysenkoism became the official state-endorsed biology.

Communist parties in other countries soon followed and endorsed this new party line. In France, the Paris communist newspaper *Les Lettres Françaises* described Lysenkoism in August 1948 as a 'great scientific event'. This debate also extended to the philosophy departments, sometimes with devastating consequences:

I also remember a guy in my class, biologist or zoologist – well, a brilliant guy – who committed suicide after a well-watered dinner during which one of the guests, who was both a professor at the Sorbonne and a member of the Communist Party central committee, had explained to him at length that the 'proletarian biology' of Michurin and Lysenko – which he taught, however – was in fact a fraud from a scientific point of view. This is the atmosphere of the *École Normale* at that time, with the blessing of Althusser. (Serres 2014, 38)

But not all French intellectuals went along (see Marks 2012). One prominent critic of Lysenkoism was the biologist Jacques Monod, who would swiftly respond in the newspaper *Combat* with an article 'La Victoire de Lyssenko n'a aucun caractère scientifique'. *Combat* was previously edited by Albert Camus, who shared a critical attitude towards Soviet politics and science, mainly due to his personal friendship with Monod (see Carroll 2013). The biologist François Jacob would similarly oppose Lysenkoism, later claiming that his decision to focus on genetics was a product of this opposition (e.g. Jacob 1981, 36; 1987, 234).

Monod was friends not only with Camus but also with Serres. Though it is hard to track down whether the story is true, the latter would tell how Monod approached him after one of his classes to ask for feedback on his new manuscript: *Le Hasard et la necessité* (1970). 'From then on we became very good friends. He introduced me to a small circle that met at his house, a kind of club where brilliant minds would meet: René Thom, François Jacob, Marco Schützenberger and a few others' (Serres 2014, 49). These friendships contrasted strongly with his experience at the philosophy departments, where 'at the very end of the 1960s my professors of philosophy were still attacking Monod, and for unsound ideological reasons' (Serres and Latour 1992, 13). A similar failure characterized his attempt to bring Monod into contact with Canguilhem:

> I even tried to introduce Monod to Canguilhem, who was after all the philosopher of the life sciences. Except that the paradigm he supported dated from the physiology of the 1940s. He had no idea what biochemistry could entail, let alone the genetic code, nor that one would soon consider deducing the totality of a living being from the DNA algorithm! He was in the past and Monod in the future. I tend to think he made me pay for this paradigm break. It must be said that such an epistemological bifurcation was difficult to swallow for a man who had dominated the discipline for so long. Anyway, he didn't want to meet Monod after all. (Serres 2014, 50)[1]

In contrast, Serres saw in Monod the embodiment of a 'new new biological spirit' (Serres 1972, 60). Similar to how in mathematics an internal epistemology is at work, Monod's molecular biology exemplified 'a "natural philosophy" intrinsic to its scientific activity' (Serres 1974a, 43). Monod indeed developed a 'ethics of knowledge', which interpreted the scientific activity as an extension to the 'noosphere' of the mechanisms at work on a molecular scale.

What most likely influenced Serres's assessment of Marxism was the famous clash between Monod and Althusser. Monod would, together with François Jacob and André Lwoff, win the Noble Prize in Physiology or Medicine in 1965 for their contributions to molecular biology. Due to the attention for this Nobel Prize, as well as the later popular and philosophical publications by these biologists, their work was picked up by several philosophers, not only Michel Serres but also Michel Foucault, François Dagognet (see Chapter 4) and Jacques Derrida (see Talcott 2014; Erdur 2018). In 1967, Monod was elected to the faculty of the Collège de France. He delivered his inaugural lecture on 3 November, a lecture that was published in full in *Le Monde* on 30 November.

In the same year, together with some of his students, Althusser started a series of lectures for a 'course in philosophy for scientists' (see Althusser 1974a; Macherey 2009). In these lectures, Althusser would respond to Monod's inaugural lecture, in which the latter offered a first sketch of his view on life and the ethics of knowledge. Although often portrayed in very antagonistic terms, recent scholarship has rather shown that Althusser was quite positive about Monod

(Turchetto 2009; Tirard 2012). Althusser indeed described Monod's lecture as 'an exceptional document, of an unparalleled scientific quality and intellectual honesty' (Althusser 1974a, 145). Though Monod was a clear critic of Marxism, Althusser welcomed his critique on teleology and his reconceptualization of life and complexity as aleatory. Althusser himself was in fact very critical of the classical conception of dialectical materialism, rather aiming to reconceptualize what it meant, making Monod rather an ally:

> Monod does not *declare* himself to be a materialist or a dialectical thinker. These words do not appear in his text. But everything he says about modern biology displays a profound materialist and dialectical tendency, visible in positive assertions coupled with determinate philosophical condemnations. (Althusser 1974a, 147)

But where this alliance breaks down, is in Monod's subsequent step, namely to extend his reflections beyond the biosphere in what Monod – following Teilhard de Chardin – calls the noosphere: the world of ideas, language, history and ethics. Monod ends his plea, therefore, with the claim that 'language created man', something that Althusser could only see as an idealist and ideological statement:

> In making this extrapolation, Monod believes himself a materialist because for him language is not a spiritual origin, but simply an accidental emergence which has the informational resources of the human central nervous system as its biophysiological support. Yet, in his theory of the noosphere Monod is in fact (though not according to his stated convictions) idealist – to be precise, mechanistic-spiritualistic. (Althusser 1974a, 150)

Althusser thus uses Monod as an illustration of the task of the Marxist philosopher: the philosopher must step in and make a demarcation between scientific practice and ideological statement, unmasking the spontaneous philosophy of the scientist. In the case of Monod, this is his illegitimate move to 'arbitrarily impose upon another science which possesses a real object, different from that of the first, the materialist content of the first science' (Althusser 1974a, 151). This dismissal of Monod was also taken up by other Althusserians (e.g. Pêcheux and Fichant 1969).

Monod, who got his hands on a written version of the lecture, was not impressed by Althusser's criticisms (Monod 1970, 40). Molecular biology and dialectical materialism were for him incommensurable. He could not but dismiss Althusser's criticisms as unscientific, and even laconically responded by a reversal of Althusser's famous notion: Althusser was merely articulating a 'spontaneous biology of the philosopher'. Michel Serres would later copy this move, dismissing attempts by philosophers to mobilize sciences for their own benefit as instances of 'the spontaneous science of the philosopher' (Serres 1974b, 46). But soon followers of Althusser came to his defense. For instance, Lecourt writes in his book on the Lysenko affair how 'these arguments [about Lysenkoism] give Monod's positions force and an audience. Not to answer them is to allow real and not imaginary motives for scientists' persistent distrust of Marxist philosophy to survive intact' (Lecourt 1976, 101).

In his own review of *Le Hasard et la nécessité*, Serres would respond to this accusation of idealism by Althusser. Serres starts by stressing how idealism has become a purely derogatory term. 'The now cursive definitions of idealism are all isomorphic to those of that what leads to error. It is therefore tautological to dogmatically demonstrate the errors of idealism' (Serres 1974a, 55). For Serres, there are only two ways to meaningfully define idealism. Either idealism entails a mathematical idealism, in the Platonic sense, or an idealism of the subject, where 'the world is nothing but my representation' (Serres 1974a, 55). Serres dismisses the second type of idealism: 'A century and a half of critique has shown, I believe definitely, that it was nothing but a mythology' (Serres 1974a, 55). But Monod is not part of this subjective idealism since a focus on the subject or representation is absent in his epistemology. Instead, to the extent that Monod is an idealist, he must be of the first Platonic type, where the ideas have an objective existence, outside of the subject. But instead of invoking a Platonic world, the ideas can be materially situated: 'Monod knows *where* its invariant form is: it is written on the DNA tape. Finally, genetics was one of the first sciences to relativize, once and for all, the activity of the individual subject' (Serres 1974a, 56).

It is very telling that it is precisely in his text where he rebuts this Althusserian accusation of Monod as 'idealism', Serres will describe Monod's work as a 'non-Bachelardian epistemology' (Serres 1974a, 57). To dismiss Bachelard was thus indirectly also a dismissal of Althusser and vice versa.

Michel Serres and Marxism

Concerning Marxism, there are two main points of critique found in the work of Serres, aimed at two central concepts of Marxism: production and class. The first critique of Serres is that a Marxist production-centred view might have become outdated. 'Breaking with the then dominant ideology of Marxism, I said that the society of communication was taking over from the society of production'

(Serres 2014, 111). For Serres, it was through his fascination with the novel concept of communication that he

> parted ways, breaking with the vulgate shared by most philosophers of the time, which was broadly speaking a Marxist one (especially with Althusser at the École Normale), and which sought to foreground problems of production. I said no, the society of tomorrow will be a society of communication and not a society of production. (Serres 2003b, 230)

This claim was also central to Serres's session of the *Société française de philosophie* in 1968. Here, similarly, Serres starts with the premise that '[u]ntil now, the technological environment has favored objects intended for production' (Serres 1968b, 34). But the locus of innovation and the economy is shifting towards communication: 'The *ars producendi* and the *ars inveniendi* depend together on an *ars communicandi*, of which it is urgent that it be the core of a philosophy' (Serres 1968b, 35). In the second part of the lecture, Serres links this theme to his general philosophy of communication, interpreting the world as one giant network (see Chapter 4).

As to be expected, other philosophers were sceptical. Serres later recounted how 'Louis Althusser, who then reigned over the École Normale and the intellectual world, was furious. He struck me down, claimed that I had become a fascist: "Michel, you have become mad! What does this mean politically"' (Serres 2014, 119)? Similarly, his supervisors, Canguilhem and Hyppolite, both had their doubts. Canguilhem raised the same question about the political implications of this view. Hyppolite, on his turn, contested the historical claim: 'Where did you get that, what is the concrete motive that led you to say that communication is becoming the centre of everything? And at the end you fall back on a transcendent intersubjectivity which does not seem to me to follow from what precedes' (Serres 1968b, 60). In his reply, Serres stresses how his claim is not so much that communication has replaced production, but that the focus has shifted towards communication:

> I only said that currently the set of productive techniques is no longer the elective place of problems, it is the set of communication techniques that is the elective place of problems; and this is the distinction that I wanted to make: not that one is dead and the other begins, but simply that the problem has shifted, that there has been a transfer. (Serres 1968b, 61)

Serres later recounted how, when meeting engineers at that time, 'they all felt that production issues were settled and that the central question now was how to communicate and how to sell.' But this was something that the Marxists did not pick up. Thus in addition to the nonsense they say about science, 'they were completely unaware of what was going on in the world of work, they did not see how one built a road or a house' (Serres 2014, 122).

A second, less prominent criticism of Marxism is found in *Genèse* (1982). As we will explore in later chapters, Serres is interested in how order comes out of disorder. This question is a question not only about the physical world but also about social order, and thus about violence and peace. 'According to theories, classical ever since the nineteenth century, social violence is to be had through class struggle. Fury here is the product of classes. Violence is secondary, classes are primary' (Serres 1982a, 84). In this context, Marxism pops up, which explain social history in terms of class struggle. Again, Althusser is the main example here, though Serres does not cite him:

> History is a process without a Subject or a Goal where the given circumstances in which 'men' act as subjects under the determination of social *relations* are the product of class struggles. History therefore does not have a Subject, in the philosophical sense of the term, but a *motor*: that very class struggle. (Althusser 1974b, 99)

As we already saw in the previous chapter, Serres himself was fascinated with the notion of classification and their origins. The same question he raised there in the context of Comte, he will raise here as well:

> Well, in sociology, as in biology, in biology as in any classification, of sciences, beings, stones or numbers, classes are not essences. Whether they're present in nature or just through our knowledge, they remain products. In other words, the class has itself a history and a makeup, the classing has itself a history. To believe that class struggle is the driving force of history is to believe that class is outside of history, it is to say that classing remains eternitary. This cannot be the case. Class is in history and so is struggle, the driving force is elsewhere. (Serres 1982a, 84)

Serres thus reverses the order: class does not create history, but history creates classes. Classes are not given originally, but are created in response to an expanding complexity. 'When a system expands, in dimension, number, and complexity, it always has a tendency to form into subsets, having all the more distinction the larger the expansion, all the more specialized thereby, all the more separate the more the system tends to maintain its cohesion' (Serres 1982a, 84). Classes are thus a product of a threshold effect, where social expansion reaches a point where disorder lures. In this, Serres will follow René Girard in arguing that there is an original and growing mimetic violence present in history (see Chapter 6). The constitution of classes is a way to contain that violence, steer it in an acceptable and sustainable pattern. 'Division into subsets minimizes fury. Division into subsets protects, preserves the unity of the body as a whole, because it tempers this free energy and channels it' (Serres 1982a, 85). Or to quote Serres more extensively:

> The division into classes is carried out under the pressure of a great danger, this danger's name is fury. I shall thus no longer say class struggle, but classes born of struggle. Try to put yourself outside of class and you will very soon see for yourself that the wind is much more violent on that plain than when sheltered by the group. I know very few exceptions to this martial law: everyone, with a shiver, snuggles back up with a pressure group, a class. And if a class were to speed up the struggle and the violence, it would very shortly be an empty class. (Serres 1982a, 85)

The legacy of an Althusserian Bachelard

We now have an idea of how Serres's relation to French epistemology must be understood in the context of a rising Althusserianism. But Latour, in a similar vein, started his career in the 1970s, at the height of the Althusserian school. To study the sciences in France during that time meant to relate oneself to authors such as Althusser, Foucault and Canguilhem, and their readings of Bachelard (Schmidgen 2015, 58). Latour was initially not completely dismissive of Althusser's work, even admitting that in the 1970s, besides his fascination with Gilles Deleuze, he 'had organized another reading group on *Lire le Capital*' (Latour 2001, 138).

Nevertheless, he soon distanced himself from Althusser's position. In fact, in some passages Latour refers to Althusser directly, describing the 'epistemological break' as 'Louis Althusser's favourite (and fully modernist) expression' (Latour 2010b, 480). It was Althusserianism that 'was going to purge the sciences from all external influence by separating them for good from the ideology' (Latour 2006, 208). And what was especially problematic was how according to 'this tradition, rationality is exercised only through a continual asceticism that separates it from what makes it exist' (Latour 1999a, 267n9). We thus find the model of purification here, but ascribed to Althusser.

That Latour was mainly struggling with Althusser and not French epistemology in general is also clear from the fact that Latour is quite positive about the work of other French epistemologists, such as Pierre Duhem or François Dagognet (see Chapter 4), which he sees as 'dissident epistemologies' (Latour 2003, 72). In some interviews he even includes Bachelard, claiming for example that 'my tradition was rather that of Duhem and Bachelard, realist and rationalist, and passionate about the number of mediations that allows us to speak the truth' (Latour 2001, 140).

The clearest case, however, is the work of Stengers, who often refers to Bachelard and Althusser together, for example criticizing their view 'that the "historical" history of the sciences is permeated by opinion, or, in Althusser's terms, by ideology' (Stengers 1993, 26). She is, in fact, quite reflexive about how it is the Althusserian reading that is at play here:

> The term epistemological break comes from Gaston Bachelard, but its extraordinary influence in French epistemology appears to be linked less to the specific content Bachelard constructed for it through examples drawn from physics or chemistry than to the strategic function it played in domains he himself did not tackle. Having become a 'cut', it allowed Louis Althusser to sanction the scientific character of Marxist theory. (Stengers 1993, 25)

It should be clear by now that to understand the specific attitudes of these authors, especially that of Michel Serres, to the work of Bachelard, the historical role of Althusserianism must be recognized. It was this Althusserian legacy that they were fighting, and their criticisms of Bachelard therefore should not reflect the whole work of Bachelard, but rather only one very narrow Althusserian reading of it. That Bachelard is not dismissed completely becomes clear by looking to their attitude towards the concept of phenomenotechnique.

Phenomenotechnique in science studies

In contrast to the notion of epistemological break, the notion of phenomenotechnique is still very popular (Rheinberger 2005; Chimisso 2008b). Especially within fields such as STS the term was taken up from the 1980s rather uncritically, to point at how scientific instruments are constructive and shape phenomena, rather than merely describing things out there. Steve Fuller, for example, even equates the epistemological rupture with 'the temporary suspension of one's natural attitude to the world that results when a new instrument of perception is introduced to the lifeworld'. He illustrates this with the example of 'the telescope, which, once it became "naturalized" in the European lifeworld, was taken to provide an access to the world as immediate as that provided by one's unaided senses. In short, the telescope became what Bachelard calls a *phenomenotechnique*' (Fuller 2002, 125).

Bachelard himself introduced this concept in his early works, around the early 1930s, in the article 'Noumène et microphysique'. He claims that within contemporary physics one has shifted away from the unity of experience as a starting point. The relevant phenomena are not to be seen as given, but rather as a (temporary) product of the instruments and theories of physics. Similar to the notion of a 'working hypothesis' he speaks of a 'working phenomenology' where 'microphysics is no longer a hypothesis between two experiences, but rather an experience between two theorems' (Bachelard 1970, 15–16). Bachelard will later make similar claims about chemistry, for instance when it started to synthesize artificial atoms and molecules that were not present in nature. He illustrates this by contemporary examples, such as technetium, an atom which was first synthesized in 1937, and did not exist in nature. Nature might stop with heavy atoms such as uranium, but science can go further, it can 'make bodies exist that do not exist' (Bachelard 1953, 22). In this sense 'the power of phenomenotechnical variation is a novel instance in philosophy. It doubles the real by the realized' (Bachelard 1953, 197).

For Bachelard it is a mistake to state that in these cases one is merely dealing with the actualization of natural powers and potentialities. 'It is in fact by a real abuse of language that one endlessly repeats that these techniques use natural forces. One would as well say that music is the use of natural noises or that wireless telephony is the use of electromagnetic waves. In fact, it comes down in all these cases to an artificial synthesis' (Bachelard 1972, 80). It is in this sense that he also makes his famous claim that 'instruments are nothing but theories materialized' (Bachelard 1934, 16). This same idea is still found in his later work, such as *Rationalisme appliqué*:

> The trajectories that allow for the separation of isotopes in the mass spectroscope do not *exist* in nature; one must produce them technically. They are reified theorems. We shall have to show that that which man *makes* by a scientific technique . . . does not exist in nature and neither does a *natural* range of *natural* phenomena. (Bachelard 1949, 103)

Thus, when Bachelard speaks of phenomenotechnique, he mainly has in mind the fact that contemporary sciences do not rely on natural, given phenomena, but rather that they are progressing by creating novel and artificial phenomena. Another form of discontinuity in Bachelard is at play here, not a discontinuity of mind but of matter: science also implies a break with ordinary phenomena.

The model we find at work in the concept of phenomenotechnique is therefore different from the purification model. The scientific practices are similarly seen as discontinuist, as something radically different from other social practices, but not because they are purified: that is something has been removed or taken away. Rather, it is the other way around: scientific practices are discontinuist with respect to other practices because *something is added, new connections are made and the artificiality is increased*. It is therefore better to speak of a *model of proliferation*: to improve science, one should not purify it from imagination or ideology, but add connections through instruments, recording devices, experimental manipulations.

In essence, it is this model that is at work in authors such as Serres, Latour and Stengers and that is opposed to the model of purification they ascribe to this Althusserian Bachelardism. It is remarkable that it was Latour's first book (written with Steve Woolgar), *Laboratory Life,* that introduced the concept of *phenomenotechnique* to science studies:

> It is not simply that phenomena depend on certain material instrumentation; rather, the phenomena are thoroughly constituted by the material setting of the laboratory. The artificial reality, which participants describe in terms of an objective entity, has in fact been constructed by the use of inscription devices. Such a reality, which Bachelard (1953) terms the 'phenomenotechnique,' takes on the appearance of a phenomenon by virtue of its construction through material techniques. (Latour and Woolgar 1979, 64)

Latour mobilized this concept to argue that a scientific fact such as the existence of TRF(H) was established in endocrinology not by purifying the mind of certain epistemological obstacles about hormones. TRF(H) could only be shown by introducing a whole network of scientific instruments, allowing it to become visible through mediations. What differentiates scientific practices is thus not an act of purification, but rather a practice where new and robust laboratory settings are constructed that allow ill-defined entities to become robust scientific facts (see Latour 1987a).

The role of interventions is thus crucial for Latour, since all scientific phenomena are defined through the actions they perform. '[T]here is no other way to define an actor other than through its action and there is no other way to define an action but by asking what other actors are modified, transformed, perturbated, or created by the character that is made the focus of attention' (Latour 1990b, 59). Precisely by the capacity of phenomena to intervene they can make scientific theories stronger, by collaborating with them and corroborating them.

Scientific practices precisely aim to construct new and artificial relations with phenomena. Interestingly, Latour refers to Bachelard's phenomenotechnique in this context to illustrate his point:

> No matter how artificial the setting, something new, independent of the setting, has to get out, or else the whole enterprise is wasted. It is because of this 'dialectic' between fact and artefact, as Bachelard puts it, that although no philosopher defends a correspondence theory of truth it is absolutely impossible to be convinced by a constructivist argument for more than three minutes. Well, say an hour, to be fair. (Latour 1990b, 64)

In later works Latour still refers to this idea, but rather under the banner of the slogan of '"Les fait sont faits": "Facts are fabricated", as Gaston Bachelard would say' (Latour 1991, 18; 1999, 127).[2]

Latour derives the relational ontology needed for this view from Serres, who similarly sees scientific practices as specific formations of relations (see Chapter 4). In *Les cinq sens* (1985), Serres links this model of proliferation with the metaphor of the maze: rather than seeing mazes as unnecessary diversions or distortions of a straight path, a maze can be productive precisely by introducing more interventions. Such 'distortions' enable one to detect more, to know more. Rather than letting all phenomena through without distinction, a maze can differentiate between phenomena by narrowing and slowing them down, allowing the observer to become aware of new, subtle differences. To quote Serres extensively:

> Let us change the discourse on method, let us optimize our journeys another way. We inherit our idea of the labyrinth from a tragic and pessimistic tradition, in which it signifies death, despair, madness. However, the maze is in fact the best model for allowing moving bodies to pass through while at the same time retracing their steps as much as possible; it gives the best odds to finite journeys with unstructured itineraries. Mazes maximize feedback. They provide a very long path within a short distance and construct the best possible matrix for completing a cycle. The best possible method for all kinds of reception, they are often to be found in sensation, whose problems they solve clearly. . . . Let us seek the best way of creating the most feedback loops possible on an unstructured and short itinerary. Mazes provide us with this maximization. (Serres 1985, 184)

Finally, within the work of Stengers you find a similar opposition between the model of purification and that of proliferation. For her, purification is an inadequate model, even 'worse than a lie: it is an insult to experimentation'

(Stengers 2006, 210). Such a model starts from the false idea that truth needs no technical assistance and that intermediaries are merely obstacles to the truth:

> What is of the order of the truth requires no artifice to impose itself. Or with the same objection: if the efficacy of a proposition requires an art of cultivation, is not the door open to relativism? What a horrible possibility! Must one not postulate that certain propositions have the power of imposing themselves by themselves, if we want to avoid the conflict of opinions and the arbitrariness of relations of force becoming an unavoidable horizon? The objection is all the more curious for coming from scientists who nonetheless know very well that a scientific interpretation can never impose itself without artifice, without experimental fabrications, the invention of which impassions them much more than 'the truth'. (Stengers 2015, 146)

She links this model of purification, 'the desperate search for that which, being 'natural' would supposedly have no need of any artifice, . . . to the hatred of the pharmakon, of that whose use implies an art' (Stengers 2015, 144). Against this model, she pleads for a model of proliferation that centres on the production of relations which enable us to detect and do anything within science. Technological mediators allow us to detect new things, but it can also be risky since they often destroy or mislead. The model of purification dismisses artifices because of this ambiguity, while the model of proliferation acknowledges this ambiguity as necessary and even productive.

The Janus head of Bachelard

As a result of the Althusserian legacy, a certain split within Bachelard was felt, created, developed or even cultivated. On the one hand you have the Bachelard of the epistemological break, a character often staged as the enemy. In contrast, the Bachelard of the phenomenotechnique is rather positively evaluated. The result is the strange situation where to prove a certain point one can mobilize Bachelard both as an ally and an adversary. But is such a split really present? Can one have the phenomenotechnique without the epistemological break? In fact, one could claim that such a split is impossible to make within the work of Bachelard. In his books both notions often pop up together, at the same moment of the history of specific sciences. The 'new scientific spirit' of which Bachelard speaks always comes down to the combination of a certain trust in a new way of thinking and a new way of using scientific instruments. Or as he adds, 'it is only by a derealisation of the ordinary experience that one can reach the realism of a scientific technique' (Bachelard 1949, 137).

A first way to understand how these two concepts are connected is to claim that phenomenotechnique must be seen as a very specific side of the new scientific spirit that Bachelard sees at work. This new scientific spirit is understood by him as constituted by the progressive freeing of the mind from the constraints of natural experience. Phenomenotechnique then boils down to a tool for such a liberation, namely the mind freeing itself from the constraints of given experience by being able to produce its own phenomena through scientific instruments. Phenomenotechnique is then merely a servant of the epistemological break. The only reason Bachelard speaks of it would be because it contributes to letting concepts develop themselves more freely within the mind. Such a reading can be found in Serres's work, when he claims that for Bachelard 'the purification of the object – the suppression of its historical metamorphism – is isomorphic to the purification of the subject – the suppression of its unconscious, dreaming and instinctive prehistory' (Serres 1972, 90n10). Phenomenotechnique, if we follow this argument, is just an elaborated trajectory to end with the same result: a purified mind that can think more freely without any obstacles.

Although such a hypothesis has its merits (see Chapter 3), it also has its limits. First of all, it seems to end up in a form of idealism, where an all-powerful mind constructs the objects of science. Bachelard himself, however, often contrasts his own position with idealist positions, partly as a reaction to his supervisor Léon Brunschvicg (see Chimisso 2008b). Against this idealism Bachelard mobilizes notions such as 'applied rationalism' or 'rational materialism' in order to stress the dialectical side of the story, where thinking and doing are intertwined. 'When one experiments, one has to reason; when one reasons, one has to experiment' (Bachelard 1934, 7).

Secondly, the notion of purification in Bachelard is at work at the level not only of the mind but also of objects. Here Bachelard's background in chemistry plays a crucial role, where purification also refers to a purification of chemical substances. In such a chemical purification materiality has a richer role to play than merely being an obstacle for thinking. The substances present themselves as forms of resistance, but also as powers to transform or to mix. In this context Bachelard criticizes the traditional notion of philosophical materialism, as a 'materialism without matter' (Bachelard 1953, 3). It is too abstract, and does not take the diverse roles that matter can play in science into account. There are in fact different kinds of materialism, such as mechanical or chemical materialism, that imply different methods and approaches. Bachelard introduces the notion of 'intermaterialism', a materialism that focuses on how knowledge can be constituted within chemistry by letting different material substances act on each other, creating novel phenomena (Bachelard 1953, 16). Materiality thus plays another, more active role. Purification here must be linked not with the mind, but with active instrumental interventions, and thus with phenomenotechnique. And it is here that Bachelard is clearly aware that such instruments have a richer role to play, besides mental purification, what we have called proliferation. The purification of substances does not merely aim at freeing the scientist from certain impressions but also aims at increasing our sensitivity to the different responses of matter, its different resistances and its different actions.

Such an idealist perspective also does not take into account Bachelard's attentiveness to the role of specialization and fragmentation in science. The fact that not all scientific practices can be reduced to one type of rationality, but that we are faced with 'regional rationalisms' (Bachelard 1949), shows that one cannot separate the rationality of science from its object. In contrast to the foregoing picture where phenomenotechnique only aims to free the mind from all obstacles, the opposite is at work here. Precisely the specific characteristics, linked to the specific field and object, are central in Bachelard's picture. Phenomenotechnique is therefore not a way to get rid of the particularities of the scientific object, but precisely allows us to articulate its specificities.

From this perspective one might almost turn the foregoing hypothesis upside down, and state that the epistemological break must be interpreted as a particular aspect of the phenomenotechnique. The latter notion then refers to a technical act of purification, where indeterminate, phenomenal chaos is purified into a refined and rigid scientific fact. The epistemological break is then part of this aspect of phenomenotechnique. But at the same time phenomenotechnique also refers to an act of proliferation, where the instrumental interventions open up new fields of phenomena to be expressed, creating new questions or domains of phenomena.

However such a reversed view does not take into account that a similar correction must be made regarding the concept of the epistemological rupture. As stated before, such a rupture implies both a vertical and horizontal dimension. This implies that such a rupture does not merely refer to a break with something external to science, such as ideology or imagination. It refers to ruptures within scientific practices as well. Bachelard in fact stresses that it is not just one single break, but a matter of 'perpetual ruptures' (Bachelard 1953, 207). It is in this context that Bachelard also uses the notion of 'rectification', referring to how a movement of revision and correction is central in the scientific practices.

'[T]he scientific spirit is essentially a rectification of knowledge, an enlargement of the frameworks of knowledge. It judges its historical past by condemning it. Its structure is the consciousness of its historical mistakes' (Bachelard 1934, 177). From this perspective the central connotation of the notion of the epistemological rupture is not that of purification, but that of *transformation*. It does not refer to a break with imagination or ideology, as the Althusserian reading suggests. Rather, its core message is negative: ruptures are needed since there is no fixed starting point or, differently put, ordinary experiences and phenomena are never enough. A rupture stresses the need of a productive act in order to form a meaningful but artificial setting in the laboratory where scientific phenomena can be articulated and subsequently form the basis of theories. In this sense, the epistemological break itself comes close to that of proliferation.

Bachelard's philosophy is thus richer than portrayed in the previous criticisms. And although not the focus of this chapter, similar claims can be made about the work of other French epistemologists, such as Georges Canguilhem. We already saw how Serres's own relation to Canguilhem experienced a radical break, leading Serres similarly to be dismissive of his work (see Chapter 1). But a link between Canguilhem and Althusser might have been a factor as well. Similarly, in Latour's criticisms of Canguilhem (e.g. Latour 1991, 92–3), Althusserianism might have played an important role. It is, for instance, no accident that Latour only focuses on *Idéologie et rationalité dans l'histoire des sciences de la vie* (1977), probably due to the notion of 'ideology' in its title and the fact that Canguilhem himself suggests that it was 'under the influence of work of Michel Foucault and Louis Althusser, [that] I introduced the concept of scientific ideology into my lectures' (Canguilhem 1977, ix).[3]

Three reasons, however, should make us sceptical towards seeing these criticisms as justifying a general dismissal of Canguilhem and historical epistemology. First of all, it is problematic to equate Canguilhem's notion of ideology with that of Althusser (see Chimisso 2015). Secondly, at other moments Latour positively appreciates the work of numerous students of Canguilhem, such as François Dagognet, Jean-Jacques Salomon and Claire Salomon-Bayet (Schmidgen 2014, 58–9). These students themselves are often very positive about Bachelard and historical epistemology, while at the same time agreeing with Serres and Latour on many central claims (see Chapter 4).

Thirdly, Canguilhem's own work goes beyond a mere endorsement of the epistemological break. For instance, in his book on the concept of the reflex, Canguilhem mobilizes the notion of phenomenotechnique to clarify the difference between the 1800 and 1850 concepts of the reflex. This shift does not consist purely in a conceptual break, but rather a material one:

> In 1850, the concept of the reflex is inscribed in books and in the laboratory, in the shape of apparatuses of exploration and demonstration, set up for it, and who would not have been without it. The reflex has stopped being only a concept, it has become a percept. It exists because it makes objects exist that make us understand. With regard to the phenomenon of which it is claimed to contain the explanation, it is no longer only phenomenological, it is also phenomenotechnical. . . . And in this respect it must be maintained that whereas the reflex in 1850 is more scientific than the 1800 concept, because it has its place marked in so many ways in the laboratory of the physiologist and because it explains a greater variety of phenomena, both provoked and given, it is not more scientific because it would be better explained, that is to say, deduced. (Canguilhem 1955, 161)

The shift shows itself by the fact that one starts to find the concept of the reflex not only in books anymore but in laboratories as well, showing itself as a useful concept to do things with, to create phenomena with. Thus in Canguilhem, similar to Bachelard, a richer perspective is present that does take the role of techniques and experimental practices in science into account (see Méthot 2013).

Purification and proliferation

If it is the case that the project of Bachelard cannot be split, does this then imply that all those double-faced scholars using one side without the other must abandon their claims? The possibility of using a crippled version of Bachelard cannot be excluded. But it is also possible to introduce a Janus-headed Bachelardism to enrich our understanding of scientific practices. Moreover, as I want to argue in this final section, such a dual face of science is already present in the work of authors such as Latour or Stengers. In the final chapter, we will see how such a perspective is also helpful to fully understand Serres's take on ecology, and its differences with that of Latour and Stengers.

It might already be clear to the reader that Bachelard's purported Janus head has an uncanny similarity with another Janus head, namely that of the sciences as described by Latour. In his early work, Latour (1987a) claims that scientific practices always have two sides: a *ready-made science* and *a science in the making*. The *ready-made science* is science as portrayed in the handbooks, namely a passive nature out there described by a fixed scientific subject here. It is often a story about the scientists who break free from their beliefs to go to the given facts of nature out there. The *science in the making*, however, is different, namely one where scientists are faced with hybrid networks of humans and non-humans, the material settings where scientists tend to recruit and construct the relevant allies within their networks.

The scientific practices thus have two faces: one is the active proliferation of connections and associations, which, once these connections are rigid enough, is subsequently purified out of view. 'The material setting both makes possible the phenomena and is required to be easily forgotten. Without the material environment of the laboratory none of the objects could be said to exist, and yet the material environment very rarely receives mention' (Latour and Woolgar 1979, 69).

In *Nous n'avons jamais été modernes* (1991) Latour generalizes these findings with the concepts of *purification* and *proliferation* (or what he also calls *translation* or *mediation*). Modernity consists in a Janus head, or as Latour states:

> The hypothesis of this essay is that the word 'modern' designates two sets of entirely different practices . . . . The first set of practices, by 'translation', creates mixtures between entirely new types of beings, hybrids of nature and culture. The second, by 'purification', creates two entirely distinct ontological zones: that of human beings on the one hand; that of nonhumans on the other. Without the first set, the practices of purification would be fruitless or pointless. Without the second, the work of translation would be slowed down, limited, or even ruled out. The first set corresponds to what I have called networks; the second to what I shall call the modern critical stance. (Latour 1991, 10–11)

The argument here is that to fully appreciate the perspective of Serres, Latour or Stengers, one has to go beyond phenomenotechnique as proliferation. In their philosophy of science a richer perspective is at work, and again this can be articulated by stressing how both sides of Bachelard are present. To tell the story about science one needs more than proliferation, one needs the model of purification as well.

In relation to this, it is a common misconception of the work of Latour that his project consists in an unmasking of the model of purification (epistemological break) as false in order to show the real face of science, namely that of proliferation (phenomenotechnique). Purification is not an illusion to be dispelled, since 'it is much more than an illusion and much less than an essence. It is a force added to others that for a long time it had the power to represent, to accelerate, or to summarize – a power that it no longer entirely holds' (Latour 1991, 40). The work of purification is not a false story science tells about itself, but is itself part of the work that is necessary to construct successful scientific facts. What allows scientific practices to function is neither purification nor proliferation on its own, but it is '[t]he link between the work of purification and the work of mediation has given birth to the moderns, but they credit only the former with their success' (Latour 1991, 41).

To retranslate this to the question of phenomenotechnique, one could say that scientific practices need two types of technologies to function. The first type is what one could call *technologies of negotiation*. An example would be the Keeling Curve, central to the research in climate change. The Keeling Curve refers to the famous graph that plots the amount of CO_2 in the earth's atmosphere for the last sixty years. The Curve articulates a field of phenomena by which we can subsequently start arguing that it is climate change that caused the increase of carbon dioxide or not. But to start that argument, we need the curve first. The reason we have this curve is due to the hard work of Charles Keeling, who has been designing and using instruments precisely to make us become sensitive to the CO_2 levels (Keeling 1998). Or as Latour puts it, '[w]ithout Charles Keeling's observatory in Mauna Loa and the instruments that detect the carbon dioxide cycle, we *would know* less' (Latour 2015a, 139). Other examples might be situated within so-called exploratory experimentation, such as the nineteenth-century instruments to detect electromagnetic effects or particle detectors at CERN (Steinle 2002; Karaca 2017). Such technologies first and foremost allow us to be affected by novel phenomena, without necessarily providing a fixed theory that can explain them. They constitute a framework in which we can start asking the right questions by firstly becoming sensitive to as much phenomena as possible, 'as bodies learning to be affected by hitherto unregistrable differences through the mediation of an artificially created set-up' (Latour 2004, 209). To do science one has to be sensitive to the responses, the feedback of the phenomena one is studying. Phenomenotechnique as proliferation refers to the capacity to become sensitive through artifices, precisely as Keeling's many instrumental settings allowed us to. 'I want more words, more controversies, more artificial settings, more instruments, so as to become sensitive to even more differences' (Latour 2004, 211–12).

However, in scientific practices there are also *technologies of control*. Becoming sensitive to certain phenomena is not enough. You will only persuade other scientists if you are able to rigorously make the phenomena affirm a specific theory. Many technological settings in scientific practices precisely have this aim. Think of examples such as Arthur Eddington's classic eclipse experiments, aiming to decide between Newton's or Einstein's theory of gravity, or the Meselson-Stahl experiment, aiming to decide whether DNA replicates by conserving the whole, the half, or none of the strands of the helix. The instrumental settings here do not aim to make us sensitive to a new field of phenomena, but rather function to control the phenomena in a very narrow way, forcing the phenomena to choose sides between theories. The goal is here to purify the plurality of actions that phenomena tend to portray into a rigorous and narrow pattern that allow these phenomena to strengthen one theory by becoming its 'reliable witness' (Stengers 1993, 167). In this sense, phenomenotechnique as purification plays a crucial role as well, namely by disciplining phenomena, allowing the scientist to become their spokesperson. This is for Stengers the core of the experimental practice: 'the invention of the power to confer on things the power of conferring on the experimenter the power to speak in their name' (Stengers 1997, 165).

Conclusion

In this chapter I tried to show how linking the concepts of phenomenotechnique and epistemological break in Bachelard's reception history helps us clarify a number of things. First of all, it helps us to fully grasp the reasons why Serres – and following him, Latour and Stengers – have been so dismissive about Bachelard. The whole idea of an epistemological break has received a bad reputation, to a great extent caused by Althusserianism, and his conflicts with, among others, Jacques Monod. Whereas this chapter, thus, focused on the broader context, in the next chapter we will delve in a more detailed comparison between Bachelard and Serres, and how they disagree concerning their normative model: a model of purification and a model of proliferation.

But, secondly, this rereading also enabled us to give a more interesting picture of scientific practices. I argued against a too hasty dismissal of the epistemological break in favour of a hypostasis of phenomenotechnique. Through a more fruitful and interesting reading of Serres, Latour and Stengers, I came to a view on science where both the model of purification and the model of proliferation can play a role. This led to the claim that within science technological interventions play two distinct roles, namely as technologies of negotiation, which aim to create new associations to become more sensitive to phenomena, and technologies of control, which aim to transform these still ambiguous phenomena into rigid scientific facts to support specific theories. Such a view goes against a tendency to overemphasize the role of proliferation and dismiss the role of purification. By reappraising the notion of the phenomenotechnique, one can however correct such an imbalance.

3

Purification as a practice of the self

Introduction

Despite the common themes highlighted in the previous two chapters, there is a crucial difference between Serres and earlier historical epistemologists, exemplified by Bachelard. We will see that this disagreement situates itself not so much on the descriptive level – about science or history – but on the level of the normative model of the scientific self.

Our starting point is a text from 1970, 'Reform and the Seven Sins' (*La Réforme et les sept péchés*), in which Serres gives a critical reading of Bachelard's book *La formation de l'esprit scientifique* (1938a). The latter aimed to develop a 'psychoanalysis of objective knowledge': a framework to map the epistemological obstacles at work in science and its history. Serres's text not only offers a pertinent criticism of the classical style of historical epistemology but provides us simultaneously with an innovative reading of the work of Bachelard and Serres's own alternative. For this, I draw on Michel Foucault's work on the care of the self. Similarly to how Foucault reevaluated Ancient philosophy as aiming not at the production of true propositions about the world but rather at a transformation of the self through a number of techniques, we can reread the work of Bachelard and Serres as two different models of how the scientific self should be constituted.

Serres's 'Reform and the Seven Sins'

The text 'Reform and the Seven Sins' consists of six sections. The first introductory session deals with the question of writing a book on stupidity (*sottise*). Two things stand out in this section. First, it plays with the duality of science and imagination in Bachelard's philosophy, where the stupid is linked to the imaginary in Bachelard. Serres's starts by reflecting on what a literary work on stupidity demands from its writer to subsequently turn to what it would mean to write a scientific book on the topic. Should Bachelard's attempt be classified as the second or could it be seen as part of the first?

Secondly, there is an implicit debate with Michel Foucault. Serres wrote reviews of Foucault's *Histoire de la folie* and *Les mots et le choses* (see Serres 1969). Central to these reviews is the question of *metalanguage*: Can one speak in a rational way about an irrational phenomenon such as madness? Serres links Foucault's ambition to Bachelard, for whom the question of metalanguage similarly came up: to speak about physics, Bachelard made use of the vocabulary of another science, namely psychoanalysis. For Foucault this option is impossible, since his concern is the history of psychology. Foucault thus uses a metalanguage that differs from the object-language. According to Serres, Foucault chooses the language of geometry and topology. His work is full of geometrical metaphors about the exclusion and separation of different spaces.

The second section of the text focuses on the title: *La formation de l'esprit scientifique,* the formation of the scientific spirit.[1] 'Formation is a good word for the epistemologist. Spirit (*esprit*) is a rather bad one, to be honest' (Serres 1970, 34). Serres does not pay a lot of attention to the second one, but merely highlights how it has been a central notion in French philosophy of science, going back to Comte: 'since the birth of the history of science and until recently, this discipline has been constantly perceived as the adventure of the human [spirit]' (Serres 1970, 34).

The greatest part of the section focuses on the concept of *formation*, which Serres finds more interesting since it 'at once cuts through the whole classification of science and is everywhere pregnant with positivity' (Serres 1970, 34). Formation has meaning in grammar, logic, biology, embryology, geology, psychology, politics and pedagogy. These different meanings do not necessarily align, since formation can refer 'to a construction, an architecture and a genesis' (Serres 1970, 35). We see here two main issues in French epistemology. First is the issue of the rationality of science: How can science be a construction without being arbitrary? Second, there is the issue of historical discontinuity: How can one structure become another one? Through his allusions, it is clear that Serres is not just referring to Bachelard here, but to Cavaillès (1938) and Canguilhem (1955) as well.

Since formation is an ambiguous notion, its meaning depends on the author. 'The Husserlian tradition of the *Krisis* privileges the geological layer, Piaget the institution of children, others social training. And Bachelard' (Serres 1970, 35)? Bachelard's case is unusual, since he uses all meanings simultaneously.

Bachelard's book starts with chapters on general knowledge and verbal obstacles (grammatical and logical formations) and ends with a chapter on the formation of children. However, one science is nonetheless missing: *physics*. Formation makes no sense as a concept in physics:

> *Preformed* physics has no physical meaning, it is *unformed*. It speaks about language, reason, life, the soul, society, the world and nature, but never of the only object of physics, namely a closed system. Yet, of closed systems in general, taken directly as object, will be born the theory of information for which all formation remains a thermal scandal, an epistemological paradox. Formation makes no sense in physics because physics shows us that it is physical nonsense. It is thus impossible to speak in a physical sense about formation in the physical sciences. Bachelard will therefore talk about pre-science in all the other languages. (Serres 1970, 36)

Bachelard ignores physics, but speaks about it abundantly in other works. His successors, however, are only concerned with either formal sciences or biological and socio-political sciences. By forgetting physics they forget the role of the object. 'The concrete object removed, idealism remains, well hidden, moreover, by a long discourse on the real. Our philosophical times are marked by the disappearance of the object' (Serres 1970, 36). It is in this lacuna that Serres will place his own philosophy, inspired by information theory (see Chapter 4).

In the third section Serres focuses on one final dimension of the concept of formation: 'there remains, in the region of norms, the moral meaning. Reform' (Serres 1970, 36). It is here that we find Serres's strongest criticism of Bachelard:

> Quite frankly, there is not a word of psychoanalysis in *La Formation*. Just listen to the language it mobilizes: it connotes a morality. The content analysis is overwhelming: it is a *Treatise of the Reform*. Of the mind, heart, soul, body, academic world. Formation of the scientific spirit, I am afraid that this means, in truth, reform of the soul desperate to reach the quintessential. Read carefully the advice of alchemic practice to the alchemists . . . and see whether the epistemologist's advice to physicists and chemists is not, perchance, the same? (Serres 1970, 37)

What is problematic in Bachelard's book are not so much his descriptive claims about science, but the underlying normative model. The next section reiterates this point, but through an analysis of the notion of an epistemological break (*coupure épistémologique*). This break concerns a moral, rather than a descriptive issue. What is at stake is not specific claims about the history of science, but the underlying discourse to characterize science. Again the Althusserian legacy plays a role here (he talks about *coupure* rather than *rupture*). For Serres, such a break never really destroys the old discourses. Bachelard, for Serres, is the exemplary case:

> Look at how, even in Bachelard himself, vocabularies of moral reform and of scientific revolution coexist, of biological transformation and of genetic mutation; look at how, backed by a traditional ethical discourse, he will consider curious 'breaks': the casting off of the old man, passage from species to species, solutions and absolutions, mutations, scholarly utopias. (Serres 1970, 40)

This brings us to the fifth section of the text, focusing on this normative model of reform. As a reading key for Bachelard's text, Serres's suggestion is simple yet radical: 'Non-science is the place of the cardinal sins, it can be mapped through pride, avarice, lust, gluttony, and sloth' (Serres 1970, 40). Bachelard thus uses the cardinal sins to characterize pre-science. Serres tries to prove this by referring to how the different chapters in Bachelard's book correspond to the separate sins: 'The *psychoanalysis of the realist* (VII) reprimands the miser, the *myth of digestion* (VIII) surprises the glutton in his sweet apathy, the *libido* (X) rummages the lustful' (Serres 1970, 40).

One could counter, however, that what is at stake here are the three psychosexual stages of psychoanalysis: the anal, oral and genital stage. But, according to Serres psychoanalysis is only at work at the surface. Here we meet again the problem of metalanguage. With what language can we speak about pre-scientific physics? Not physics itself, since it lacks the notion of formation. Bachelard therefore picks a human science, psychoanalysis. However, the problem is that psychoanalysis itself is a pre-science, an archaic enterprise, inspired by the same morality as the alchemy it tries to describe. 'It turns out that no gap whatsoever exists between Bachelard's psychoanalysis and pre-scientific morality' (Serres 1970, 41).

One issue remains, which Serres discusses in the sixth section: only five out of the seven cardinal sins are at work in the pre-scientific mind. What is missing is envy and wrath. However, this does not prove that Serres's reading is mistaken, since envy and wrath are present, but as the virtues of science itself: 'The new priests still thunder like Jupiter at the top of the hierarchy; they delight in terror, are obsessed with degrees, and color their competition with the greenish yellow of jealousy' (Serres 1970, 44). Here we find another central topic of Serres: the idea that science is based on violence and destruction. 'The original sin of destruction is at the origin and foundation of our knowledge' (Serres 1970, 44). The problem with Bachelard is not so much that his model is empirically inadequate, but that it risks ignoring this violence, which shows itself, above all, in the atomic bomb:

> *And now, under penalty of death, we are forced to outline a more archaic prehistory than that of Bachelard, to purify the sources of science poisoned, from the beginning, by terror.* It is no longer a question of the fine list of capital sins, but of our collective survival in the face of capital punishment. The purity of the soul is child's play in face of this risk. (Emphasis in original) (Serres 1970, 45)

We will return to this archaic prehistory, this anthropology of science, in Chapter 6. Here, I want to focus on Serres's diagnosis of Bachelard: how it offers us a novel reading, both of historical epistemology and of Serres's own ethos of science.

Gaston Bachelard and the formation of the scientific spirit

Serres's diagnosis can help us to understand why Bachelard, despite being a central and founding figure of historical epistemology, is actually an enigma in this tradition. I want to argue that Bachelard is an enigma for at least three reasons.

First, Bachelard did not produce any 'proper' historical studies of science, with the possible exception of his secondary thesis on the history of the problem of heat (Bachelard 1927). Of course, he mobilizes elements from the history of science in his work, but in a way that nowadays seems deeply problematic. Bachelard always rereads the past from the present, separating history of science from other forms of history. Contemporary historians of science are thus often amazed about the – in their view – naïve and simply problematic assumptions of Bachelard. First enigma: a founder of a tradition in history of science that did not write any respectable history of science.

Secondly, Bachelard simultaneously wrote books such as *La psychanalyse du feu* (1938b) or *La poétique de l'espace* (1957). Especially for Anglo-American readers it often comes as a surprise that Bachelard wrote on science at all, since they are typically only familiar with his work on imagination and poetry (Simons et al. 2019). Hence the debate how these two oeuvres relate to one another: How to reconcile the day and the night (Lecourt 1974), science and poetry? Second enigma: a founder who spends half its oeuvre writing about poetry and imagination.

Finally, there is Bachelard's reception history. Whereas his own work seems hardly political, his work has deeply influenced a generation of French political philosophers, ranging from Michel Foucault to Pierre Bourdieu. More particularly, Bachelard was very influential in Althusserian Marxism (see Chapter 2). Third enigma: How can an apolitical philosopher of science become a beacon for political philosophy?

Serres's interpretation offers a solution to these enigmas. Bachelard's work should not be read as a description of the actual history of science, but as a normative project: the outline of an ethos, a treatise on the reform of the scientific spirit. What is central in Bachelard's project is not history, but pedagogical formation. The importance of pedagogy for Bachelard has been noted before (Chimisso 2001; Wunenburger 2013). Chimisso correctly stresses the importance of Bachelard's experience as a physics and chemistry teacher. 'His epistemology and his pedagogy were not separate: rather, his epistemology shaped his pedagogy, and his pedagogy inspired his conception of science' (Chimisso 2001, 73). I want to radicalize this reading. Instead of just noting the similarities, alliances and reciprocities, I want to put pedagogy at the centre, even going so far as to speculate that Bachelard's epistemology is instrumental for his pedagogy. But for this, I need to introduce Foucault's notion of the care of the self.

Foucault and the care of the self

I will restrict myself here to the later Foucault, his work on ethics (Elden 2016). Whereas his early work emphasized knowledge and power, his later work focuses on how subjectivity is formed. Of course, subjectivity was a theme in his earlier work as well, but in his final decade it becomes a question of how individuals shape their own subjectivity (Cremonesi et al. 2016). The aim is to sketch 'the history of how an individual acts upon himself, in the technology of self' (Foucault 1988, 19). He defines this type of technology as follows:

> techniques which permit individuals to effect, by their own means, a certain number of operations on their own bodies, on their own souls, on their own thoughts, on their own conduct, and this in a manner so as to transform themselves, modify themselves, or to attain a certain state of perfection, of happiness, of purity, of supernatural power, and so on. Let's call this kind of techniques a 'technique' or 'technology of the self'. (Foucault 2015, 25)

Foucault focuses on two contexts of such a care of the self: (1) the Greco-Roman philosophy, ranging from Plato to Roman authors such as Seneca (Foucault 1984b, c); and (2) early Christian authors, often in the Roman Empire (Foucault 2018). He uses this notion to reread the history of Ancient philosophy: philosophy is not so much concerned with producing the ultimate truth about the structure of the world, but is preoccupied with the 'care of the self': a range of practices that mobilize these technologies of the self to transform and improve the self. He similarly makes a distinction between *philosophy* and *spirituality*. Whereas philosophy then refers to 'the form of thought that asks what it is that enables the subject to have access to the truth and which attempts to determine the conditions and limits of the subject's access to the truth', spirituality refers to

> the set of these researches, practices, and experiences, which may be purifications, ascetic exercises, renunciations, conversions of looking, modifications of existence, etc., which are, not for knowledge but for the subject, for the subject's very being, the price to be paid for access to the truth. (Foucault 2005, 15)

Foucault suggests that we can analyse these practices through four aspects (Foucault 1984a, 352–5). First of all, there is the ethical *substance*: the object, the part of ourselves, of our behaviour, on which the individual acts upon. Secondly, there is the *mode of subjection*: the manner in which individuals are invited to take up this moral task to work on their self. A third aspect is the *self-forming activity*: the means by which individuals can ethically change their self. Finally, there is the *telos*: 'Which is the kind of being to which we aspire when we behave in a moral way' (Foucault 1984a, 355)?

Let this suffice as a brief summary of Foucault. One issue remains: if we want to apply this framework to Bachelard and Serres, one must address the question of whether a tradition of philosophy as spirituality still exists in the twentieth century. Foucault himself, for instance, suggests that this tradition has more or less disappeared. Whereas until the sixteenth century the idea remained dominant, this changes through what Foucault calls the 'Cartesian moment' (Foucault 2005, 14), which breaks the relation between asceticism and access to truth:

> Thus, I can be immoral and know the truth. I believe that this is an idea which, more or less explicitly, was rejected by all previous culture. Before Descartes, one could not be impure, immoral, and know the truth. With Descartes, direct evidence is enough. After Descartes, we have a non-ascetic subject of knowledge. This change makes possible the institutionalization of modern science. (Foucault 1984a, 372)

There are a number of replies to this issue. First of all, it is possible to contest Foucault's claim that this dimension has disappeared (see Hadot 2002, 263–4). Descartes himself, for instance, still wrote *Meditations* on how to prepare the subject to access the truth and Spinoza wrote a *Treatise on the Improvement of Understanding*. Foucault recognizes this (e.g. Foucault 2005, 47), which explains why he could also claim that one can find elements of this spiritual tradition in later authors, such as Friedrich Nietzsche or Jacques Lacan (Foucault 2005, 189).

An alternative is to interpret Foucault's claims as an invitation to reread the history of philosophy through this new lens: 'maybe the history of European philosophy from the sixteenth century should not be seen as a series of doctrines which undertake to say what is true or false concerning politics, or science, or morality' (Foucault 2010, 349). Instead we should read it a history of different ethical projects, that each proposes a specific practice to take care of the self.

A second response is to simply state that my goal is not to be completely faithful to Foucault (for this, see McGushin 2007; Cremonesi et al. 2016; Elden 2016). Instead, I merely wish to use some elements of Foucault to shed a new light on the work of Bachelard and Serres. The specifics of Foucault's ideas therefore are not that important, and certainly not dogma. I could have used other authors who have developed similar views on the history of philosophy, such as Pierre Hadot (2002) or Stanley Cavell (Lorenzini 2015).

Finally, others have mobilized Foucault's work to analyse certain elements of the scientific practice (Rose 1998; Goldstein 2005; Jones 2006; Daston and Galison 2007; Savoia 2014), ranging from the history of note-taking (Daston 2004) to Rorschach tests (Galison 2004). Nonetheless, my concern remains slightly different: it is not about an interpretation of the actual practices of scientists, but rather about the philosophies of science of individual philosophers. It is important to keep that restriction in mind.

Bachelard and the scientific self

Bachelard's work, like other historical epistemologists, is characterized by a specific form of presentism: a rereading of the history of science through present science. Though it is possible to defend this presentism on historiographical grounds (Loison 2016; Simons 2019; Vagelli 2019), an alternative is to follow Serres and Foucault, and argue that Bachelard's main concern is not historiography, but the *formation of the scientific spirit*. The question is not to understand the history of science per se, but to achieve an '*aesthetics of the intellect*' (Bachelard 1938a, 10). History of science is merely a concern because it can teach us a lesson about how the mind progresses and transforms itself.

It is in this context that Bachelard also makes his famous distinction between *histoire perimée* and *histoire sanctionée* (Bachelard 1951, 25): between the elements of the history of science that are still seen 'at work' in current thought and those elements that are obsolete and dysfunctional. One has to judge in history of science, because a dynamism of scientific concepts is an essential pedagogical factor in the development of scientific culture: a well-constituted scientific self is capable to revise, to renew its thoughts.

This link with pedagogy has led some to see this as another similarity with the work of authors such as Thomas Kuhn (e.g. Smith 2018). Kuhn similarly stresses the pedagogical role of history of science in science textbooks, but I believe in an opposite way (see also Simons 2017). The reason why Kuhn is concerned with pedagogy is that he sees it as a threat to historical accuracy: 'the textbook tendency to make the development of science linear hides a process that lies at the heart of the most significant episodes of scientific development' (Kuhn 1962, 140).

In French epistemology the model is different. Existing pedagogies of science are misleading, not due to historical inaccuracy, but because they fail to properly prepare the new pupil. Faced with new scientific phenomena, pupils risk overlooking the necessary struggles to make sense of them. Nature as presented in the classroom is already a disciplined or rationalized nature, not nature in all its complexities. Students get the wrong impression that the scientific way of thinking is self-evident, one only has to look at nature. The dominant pedagogy thus leads them to a naïve form of empiricism. The problem with history of science is thus a false image of simplicity, ignoring the struggles and reforms the mind has to go through to make sense of novel scientific phenomena:

> The history of science is often misleading in this regard. It almost never restores the obscurities of thought. It cannot therefore fully grasp rationality in the making. Our present knowledge sheds light on the past of scientific thought so vividly that we take all glimmers for lights. And so come to believe in a reason as [already] constituted before the effort of rationality has been done. (Bachelard 1949, 9)

In that sense, we can also find in Bachelard a criticism of presentism, though with an opposite conclusion: history of science should be revised *because* we need another pedagogy. We need to 'write the history of the mind' (Chimisso 2008a), but in order to produce the adequate scientific self.

The dynamic and autonomous self

What kind of scientific self should be constituted according to Bachelard? We ask here for the *telos* of this care of the self: What kind of being do we have to aspire to be? Since it concerns science, one would expect the telos to be linked to truth: we have to aspire to a scientific self that has access to the truth. This is not the model we find in Bachelard. The main *telos* is rather what we encountered in the first chapter: *open rationalism*. The scientific self has to be a self that is dynamic: a self which is in motion, constantly revising and changing itself.

Secondly, what is the *substance* of this intervention? What must be made dynamic? For Bachelard the substance is the human spirit. You find metaphors for this all over his work, especially in *La formation de l'esprit scientifique*: we have to start the autogenetic process of thought (1938a, 13); the spirit has to be transformed into one that suffers if it does not mutate (1938a, 16); we have to bring thought in a state of permanent mobilization (1938a, 18); we have to bring reason in a permanent state of being born anew, an *état naissant* (1938a, 48).

But what hinders this dynamism? And what should the subject do about it? What is the *mode of subjection*? Again the answer is simple: *epistemological obstacles*. But where do they originate from? For Bachelard, they originate from life: spontaneously, naturally given, uncultivated desires. These desires make thought stop for two main reasons. First of all, they imply *passivity*. A recurrent theme in Bachelard's work are descriptions of how one gets carried away by an image, how images are too strong and occupy one's whole thought, making any attempt for further explanation disappear (Bachelard 1938b, 73). In *Psychanalyse du feu* the central example is that of fire: it is a phenomenon that carries us away and never leads to a genuine scientific description. It is a phenomenon that immediately forces itself on us and links itself to biological desires and pre-existing images. This passivity, for Bachelard, also shows itself in the failure of the self to demarcate the problem one is dealing with. The pre-scientific self gives no clear demarcation, but is carried along by the study object in all directions, overwhelmed by all its details (1938b, 71–3).

The second issue is that these desires are *external* to the realm of knowledge. Bachelard famously claims that 'opinion does not think' (1938a, 16). It does not think, because it does not serve the purpose of knowledge, but other vital purposes. To break free from this passivity and exteriority, the scientific self has to be freed from life. Bachelard describes this in numerous ways: we have to destroy our primary will in order to constitute a new, second will (1936, 14); pupils come into our classrooms not to be cultivated for the first time, but to change cultures (1938a, 16); we have to learn to unlearn (1934, 86).

What then, if not life, is then the engine of the proper dynamism of the scientific self? For Bachelard it is thought that moves itself: instead of passively being moved by images, one moves oneself. We will see that what gives thought this autonomy is an object of debate: Is it the will that moves itself?

Is it the faculty of judgement? Or do concepts possess their own normative movement? This autonomous movement reverses the two reasons why life was problematic. First of all it is *active*, not passive. It is about the organizing capacity of our spirit (1936, 18). In *Rationalisme appliqué* (1949, 38) Bachelard gives the following example: though we see, for instance, that leaves do not fall in a straight line to the ground, we can abstract from the 'accident' of friction and see the motion of free fall. Secondly, this dynamism has its *ground in itself*: it does not respond to external questions, but to the question posed by itself (1938a, 23). Here, Bachelard is inspired by Pierre Janet's psychological study of what it means to start an action, to start a thought, to begin something new (Bachelard 1936, 4).

Let me give some further characteristics of this autonomy. A first theme is that of the need to *purify* oneself from this life: abstract thinking is a form of asceticism (1938a, 283). A common metaphor here is that of renewal: the scientific self thinks anew, renews itself. Here we find the metaphor of the new versus the old man: the human spirit appears in front of science not as a young mind, but exactly already as an old man, occupied by a complex network of existing desires and images. The task of the scientific spirit is to rejuvenate, to make young again what is old (1938a, 8).

A second theme is that of immobilization. The problem is not to get something moving which did not move at all, but to transform something that moves passively into something that moves actively. For this, the self first has to be stopped. The scientific self has to install a break on the original desires. The self has to train itself in scientific patience (1938a, 8); the self has to install a philosophy of repose, to empty the full life (1936, v); the self has to install an internal brake that stops, calms down and breaks to a halt (1936, 19).

Thirdly, this struggle shows itself in an internal struggle in the self: the true self is not to be found in ourselves, but the self rather has to be dismantled and destroyed and rebuilt in a new and better way. In order to access knowledge, the self has to desubjectivize and dismiss its own subjectivity (1938a, 294); the self has to think against one's own brain (1938a, 299).

Fourthly, this struggle does not occur only once, but repeats itself endlessly. There is the constant risk of falling back, of letting images slip in (1938a, 8). It is a struggle that reappears whenever the self encounters a new object of scientific study (1949, 15). It is in this context that we find another famous quote by Bachelard: 'Are we rationalists? We try to become so' (Bachelard 1942, 24). Rationalism is not a state the self has to achieve only once, but repeatedly, by remaking the rupture with life again and again.

The endlessness of this struggle also partly explains why Bachelard does not shy away from mixing different historical periods when describing the pre-scientific mind. Chronology is irrelevant, because with each new phenomenon, the self is faced with the same struggle: to bring the dynamism of life to an end and replace it with the autonomous movement of scientific thought. The constitution of the scientific self 'cannot be treated on the historical plan because the conditions that led to *rêverie* in the past have not been eliminated by contemporary scientific education. Even the scientist, when not practicing his specialty, returns to the primitive scale of values' (1938b, 4).

The scientific self

What is the specific element that for Bachelard grants the self its autonomous dynamism? We will see that this is a complex matter, but the answer is along the lines of the scientific *judgement*, or the scientific *concept*. What is characteristic of this model is that concepts (and mathematics) are more than a language to express thoughts. Instead, they play a more active role. More specifically, this implies four aspects of concepts.

First of all, a concept is understood as a *judgement* (Bachelard 1936, 16). To understand a concept is to *remake* the implicit judgement. Here a number of issues are at play. First of all, this understanding of concepts can be traced back to the influence of the French tradition of reflexive analysis (*analyse réflexive*). Represented by Jules Lachelier, Jules Lagneau and Léon Brunschvicg, this movement was the particular way in which Kantianism was taken up in France, linked to a broader movement called French spiritualism (with Victor Cousin and Félix Ravaisson). Lachelier and Lagneau articulated this programme of reflexive analysis as a study of the essential act of thinking: the faculty of judgement. Here, a judgement is understood as the creation of a relation between two separate ideas. A judgement is thus a creative activity in the world rather than a mere description of that world.

Secondly, a concept is *relational* and *discursive*: one is not faced with one concept in isolation, but with a field of concepts (Bachelard 1938a, 123, 154). The philosopher Alain, linked to this tradition as well, summarizes the thought of Lagneau as follows: 'The idea of Thought is the idea of the necessary implication of everything in everything, the idea of the dependence of each Thought on all the others. It is therefore necessary that, beneath each Thought, all the others are always implicitly present' (Chartier 1898, 537). Thirdly, since a concept consists in the implicit judgement understood as an activity of creating new relations in the network of ideas, there is no endpoint. Concepts are open-ended, always referring to new ideas, new consequences and so on. They often contrasted this view of concepts as active judgements with a logical view which focuses on logical propositions, which French scholars saw as nothing but fossilized versions of the dynamic process of thought that preceded them.

Neither Lachelier nor Lagneau really engaged with contemporary science. Brunschvicg, however, wanted to update Kantianism in the light of recent scientific developments, such as Einstein's theory of relativity. He saw his own work as a kind of unification of the reflexive analysis school, history of science and the novel insights from anthropology and psychology. In several places Brunschvicg applauds 'the happy synchronism' which 'connects the reflexive analysis of Jules Lachelier or of Lagneau with the criticism of the primitive, puerile or pathological mentality, such as it results from the work of a Lévy-Bruhl, of a Jean Piaget, of a Pierre Janet, to refer only to the leaders of these schools' (Brunschvicg 1931, 196).

A third aspect of concepts is their social character. We find this theme in Bachelard's characterization of psychology as *interpsychology* and of rationalism as *interrationalism*, always involving a relation to the other (Bachelard 1949, 19–21), and moves from *cogito* to *cogitamus* (1949, 57). But most clearly it is found in Bachelard's notion of the scientific city (*cité scientifique*). For Bachelard, science is a collective enterprise, because an adequate pedagogy implies social relations. To understand is to be able to teach the other, including oneself. There is thus the necessity of a scientific city aimed to educate one another: the scientific self is a permanent pupil.

Finally, the concept is characterized by an element of polemics, negation and battle. Since scientific thought is constituted through an epistemological rupture with life, it is in essence defined by negation. This negation not only is the starting point of scientific thought but also characterizes the continuous process that occurs within science, exemplified by scientific revolutions. Knowledge is always produced *against* previous knowledge (Bachelard 1938a, 13).

Though Bachelard is part of this psychologistic programme of reflexive analysis, his work nonetheless has additional elements that complicate this picture. First of all, Bachelard focuses on the autonomy and internal normativity of concepts and mathematics. This anti-subjective dimension will be taken up by later epistemologists, such as Canguilhem, Althusser or Foucault. Secondly, Bachelard stresses the role of experiments and instruments, embodied by the notion of *phenomenotechnique* (see Chapter 2). We saw how Serres suggested that phenomenotechnique could be read in function of this ethical project: 'the purification of the object – the suppression of its historical metamorphism – is isomorphic to the purification of the subject – the suppression of its unconscious, dreaming and instinctive prehistory' (Serres 1972, 90). The self needs to produce its own phenomena, rather than take them as naturally given, in order to become more autonomous and dynamic. This second, technological dimension was taken up by scholars such as Latour or Rheinberger, often without the normative project.

The poetic self

We already encountered the enigma of the two oeuvres in Bachelard: science and imagination. Within the secondary literature this is a main topic of concern: How to reconcile both oeuvres? The most traditional way to grasp this is through the contrast between day and night: these are two aspects of human life, but different and incompatible ones (Lecourt 1974). But this dualist scheme is misleading, since it equates imagination and poetry to natural desires. It is more helpful to see it as a threefold scheme: science *and* poetry against life. We thus find here a second model of an autonomous dynamism, namely the poetic self. Whereas in early works Bachelard still take these two faculties together, opposed to life, in later works he treats them separately:

> One must love the psychic forces of two different types of love if one loves concepts and images, the masculine and feminine poles of the Psyche. I understood that too late. Too late, I found a clear conscience in alternating work between images and concepts, two clear consciences, one belonging to broad daylight, the other accepting the nocturnal side of the soul. For me to be able to enjoy a double clear conscience, the clear conscience of my double nature finally recognized, I would have to write two more books: a book on applied rationalism and a book on active imagination. (Bachelard 1960, 53)

The poetic self functions analogously to the scientific self. First of all, the poetic self is not to be identified with life, but in opposition to it. This is clearest in *La dialectique de la durée*, which includes a chapter on how music and poetry create a new rhythm that goes against the rhythm of life itself. Secondly, the poetic self has its own dynamism. The poetic is the capacity to bring petrified metaphors and images back into movement. In that sense imagination is first of all 'the faculty of *deforming* the images offered by perception, of freeing ourselves from the immediate images; it is especially the faculty of changing images. If there is not a *changing* of images, an unexpected union of images, there is no imagination, no *imaginative action*' (Bachelard 1943, 7). In the poetic sphere, this dynamism is not self-evident, but needs to be guarded and protected: but not only from life itself but also from concepts. Bachelard warns us not to use concepts to explain images: 'By giving stability to the image, the concept would stifle its life' (Bachelard 1960, 52). Instead, the 'image can only be studied through the image' (Bachelard 1960, 53).

It is from this perspective that Bachelard creates a whole project of literary criticism, focused on the images linked to the four elements (fire, water, earth, wind) and focused on finding the dominant element in the oeuvre of specific authors, for example air in the case of Friedrich Nietzsche (Bachelard 1943). Moreover, there is once again a particular role for materialism in the imagination, a material imagination focused also on the movement of the body and the hand (see Rheinberger 2016).

The self-forming activities of the scientific spirit

We already encountered three dimensions of Foucault's care of the self. The *ethical substance*: the mind, the psychological thoughts and desires that have to be reworked. The *mode of subjection*: one has to constitute the dynamism of the self against vital desires. The *telos:* a dynamic and autonomous self, that is free from the passivity and externality of life. But what about the *self-forming activity*: 'the means by which individuals can ethically change their self' (Foucault 1984a, 355)? What are the specific technologies of the self that Bachelard prescribes?

We can find a first one in La dialectique de la durée (1936): *rhythmanalysis*. 'Rhythmanalysis' is a term Bachelard borrows from the Portuguese philosopher Lúcio Alberto Pinheiro dos Santos, an enigmatic figure whose manuscript is unfindable, raising doubts whether he even really existed. In the meantime, his existence has been ascertained, and his book was probably written, but lost. Rhythmanalysis is used in response to Bergson's model of duration: instead of accepting the natural rhythms of life, one has to install a new rhythm, which enables autonomy and dynamism. It is thus a practice to investigate which rhythms, which durations shape one's self, and to replace them with better rhythms. A specific example Bachelard gives is that of a tooth ache:

> With a little practice, we can for example make toothache vibrate. All we need do is calmly and attentively put it into its proper perspective and avoid the general annoyance and agitation that would fill up the intervals of the particular pain. The throbbing of this local pain then acquires its regular rhythm. Once this regularity has been accepted, it comes as a relief. Pain is truly restored to its local aspect because its correct temporal aspect has been fully determined. (Bachelard 1936, 146)

Bachelard gives other examples, such as breathing exercises, installing a daily rhythm in your work, linking it even to Indian mysticism. But in contrast to

Pinheiro dos Santos Bachelard is mainly concerned with the rhythms of thought, the scientific self.

The second example is *psychoanalysis*, not of the subject, but of objective knowledge. Bachelard speaks of a reflective psychology: we focus on how human desires are reflected in its objects of study, and the accompanying images (Bachelard 1938a, 159). These epistemological obstacles, therefore, must not be refuted, but their origin must be uncovered. We have to examine the intentionality behind them: Is it one focused on concepts or is it inspired by vital desires? In the pre-scientific mind, the intentionalities are mixed: we think we are scientifically oriented towards the object, but in reality we are mainly driven by practical needs and lively images found in the other realms of life (Bachelard 1938a, 169). In *Psychanalyse du feu* Bachelard proposes a taxonomy of complexes at work behind these images, for example the Prometheus complex: the desire to know more than our father, superiors and so on (1938b, 20).

The final example is found in *Philosophie du non* (1940): *epistemological profiles*. As we saw in the first chapter, whereas philosophy tends to be closed in its categories, the new scientific spirit demands from us an open philosophy. One way Bachelard grasps this openness is through a 'differential philosophy' (1940, 14). We need a differential philosophy, according to Bachelard, to analyse the consistency of our scientific thoughts, almost in the way a chemist analyses a substance to see which components are present and dominant.

Every scientific concept is associated not with one philosophy but with several, going through different stages, depending on the context and the maturity of the scientific spirit. Bachelard presses the point, asking rhetorically: 'Do you really believe that the scientist is a realist in all his thoughts? Is he realist when he assumes? Is he a realist when he resumes? Is he a realist when he schematizes? Is he a realist when he is mistaken? Is he necessarily a realist when he affirms' (Bachelard 1940, 41)? For Bachelard the answer is no: realism is merely one stage of the dynamism of the scientific spirit.

Nonetheless, concepts can get stuck in one stage. Take the concept of mass, that is initially faced with the obstacle that it is confounded with size: 'For a greedy child the larger fruit is the best, the one that speaks most clearly to his desire, the one that is the substantial object of the desire. The notion of mass embodies the very desire to eat' (Bachelard 1940, 22). To relieve the scientific spirit from such obstacles, one has to draw up epistemological profiles for all one's basic concepts. 'It is by such a mental profile that one could measure the effective psychological action of the diverse philosophies in the labor of knowledge' (1940, 42).

Science and ideology in Louis Althusser

Let me briefly end, stressing how this model of Bachelard, though influential in France, also underwent crucial transformations (as we already saw in the previous chapter). Around the Second World War, this psychologistic framework is abandoned in favour of what one could call an anti-humanist programme, typically linked with the name of Jean Cavaillès (Cassou-Noguès and Gillot 2009). Cavaillès was a student of Brunschvicg, mainly focused on philosophy of mathematics, and killed during the war for resistance activities. His name became associated with one particular philosophy which served as an alternative to subject-centred philosophies: Spinozism.

This story surrounding Cavaillès and Spinozism is perhaps more myth than reality, created by a number of French philosophers (Granger 1947; Canguilhem 1984; Foucault 1989). The myth is best expressed by the book of Knox Peden: *Spinoza contra Phenomenology* (2014). It is the myth that Spinozism was the only viable alternative against German phenomenology. And within this myth Cavaillès played the crucial role of replacing a philosophy of the subject with a philosophy of the concept: 'It is not a philosophy of consciousness but a philosophy of concepts that can provide a theory of science' (Cavaillès 1947, 78). Foucault most famously propagated the myth that in France one had to pick sides concerning phenomenology, either a philosophy of the subject or a philosophy of the concept: 'one network is that of Sartre and Merleau-Ponty; and then another is that of Cavailles, Bachelard and Canguilhem' (Foucault 1989, 8).

One of the most influential forms in which the new Spinozist programme was developed is, once again, Althusserianism. In that sense, the latter takes up a number of normative questions we explored earlier, but transforms it in a number of significant ways. Similar to Bachelard, the goal is to open up, to clear the road for scientific practices to become dynamic again. The philosopher's action, which 'divides the scientific from the ideological[,] has as its practical effect the 'opening of a way', therefore the removal of obstacles, opening a space for a "correct line" *for the practices* that are at stake in philosophical Theses' (Althusser 1974a, 100). It is obvious that Althusser's reconceptualization is in Marxist terminology. The problem is transformed into one of the exploitation of the sciences by philosophy. The aim, therefore is to shift 'the balance of power within the [Spontaneous Philosophy of the Scientist], in such a way that scientific practice is no longer exploited by philosophy, but served by it' (Althusser 1974a, 141).

Secondly, the main threat to this openness is not so much life and imagination but ideology. But similar to Bachelard's model, ideology is problematic because it postulates a passive and external relation. 'An ideological proposition is a proposition that, while it is the symptom of a reality other than that of which it speaks, is a false proposition to the extent that it concerns the object of which it speaks' (Althusser 1974a, 79). Such a proposition is serving specific practical and ideological goals. For instance, one might present the claim that we live in a meritocracy as a scientific claim, but one can wonder whether it is not an ideological notion that serves, not so much the production of knowledge, but the dominant ideologies and inequalities. The aim therefore is, once again, *purification*: we have to purify the spontaneous philosophy of the scientists from its idealist and ideological elements, replacing it instead with a materialist and Marxist philosophy, which according to Althusser does not exploit, but serve the sciences (for reasons I will not go into here).

We thus find a very similar model to that of Bachelard, but with three main differences. First of all, Althusser is not so much concerned with the individual ethos of the scientist, but with the social organization of science. It is not so much a matter of philosophy as individual *therapy*, but rather as *collective emancipation*. Secondly, the model of Althusser is historical, whereas Bachelard's model seems atemporal. Whereas Bachelard naturalizes the problem of epistemological obstacles, Althusser politicizes this question: the spontaneous obstacles are products of specific political ideologies. Finally, ideology for Althusser is understood in a very materialist way, rather than purely a matter of ideas and images. Ideology is first and foremost a matter of rituals, practices, institutions and infrastructures that shape our conduct, and only secondarily our thoughts in function of the dominant ideologies (see Althusser 1995).

Michel Serres and the formation of the scientific body

In 'Reform and the Seven Sins', Serres dismisses this Bachelardian model of the scientific self. The danger of collective destruction, in the form of the atomic bomb, is more pressing than finding the correct ethos for the scientist. We will come back to this political danger in Chapter 6. However, Serres does not completely dismiss the idea of a scientific ethos. The problem is the specificity of Bachelard's ethos, not the project of a scientific ethos per se. In fact, in Serres we find not so much the absence of an ethos, but an alternative model of the scientific self. I will therefore briefly end by putting a number of themes on the agenda, which will be explored further in subsequent chapters.

The relational self

The *telos* of Serres's model of the self is very similar to that of Bachelard: the scientific self has to be transformed into a dynamic self. However, the *substance* is radically different: whereas for Bachelard one had to make the spirit dynamic, in the case of Serres it is the body that has to be made dynamic.

For instance, Serres often speaks about the self through bodily metaphors. In *Le Tiers-instruit* Serres argues that 'the self is a mixed body: studded, spotted, zebrine, tigroid, shimmering, spotted like an ocelot, whose life must be its business' (Serres 1991a, 145). Similarly, in *Petite Poucette* (2012) the issue of a new pedagogy in our digital times is framed as an issue of bodies: Will our bodies still be able to sit still in class, now that they have access to all knowledge on their mobile phones? According to Serres our body will adapt, extend itself with new tools, such as phones or computers. It is in this context also that Serres makes the comparison between teenagers and Saint Denis of Paris, who, after being executed, took his severed head under his arm and walked a few more miles to the place where he wanted to be buried: 'Not long ago, we all became like St. Denis. Our intelligent head has been externalized outside our skeletal and neuronal head' (Serres 2012, 19).

But the clearest eulogy to the body is *Variations sur le corps* (1999), where Serres strongly opposes the self to thought and thinking, and identifies it with the body:

> I have never been able to express the ego or describe consciousness. The more I think, the less I am; the more I am, the less I think and the less I act. I do not see myself as a subject, a silly project; one that meets only things and other subjects. Among them, a little less a thing and much less another subject, is my body. (Serres 1999, 12)

As a result, the kind of self to be developed is different. Instead of the ideal of returning to dynamic thought, Serres speaks of the ideal of the 'nascent body [*Corps à l'état naissant*]' (Serres 1999, 8). The exemplary figure is not the professor, the scientist, but the gym teacher:

> No seated teacher taught me productive work, the only ones that did succeed in this, my gymnastics masters, coaches, and later guides, inscribed the conditions [for productive work] in my muscles and bones. They teach what the body is

capable of. Do you want to write, research, enter into a life of work? Follow their advice and their example. (Serres 1999, 44)

Similar to the substance, Serres's *mode of subjection* is quite different. It is a matter of teaching the body to be affected and sensitive in novel ways (see Chapter 8). Returning to a nascent body is making this body transparent, capable of detecting new and subtle differences in the world. 'If the mind is born from the transparency of the body, from its faculty or possibility of doing, the very first ideas must emerge from this virtuality, showing these same characteristics' (Serres 1999, 124). In that sense, to learn implies not so much thought, but the absence of thought: 'To better inhabit your body, as well to better give it orders, forget about it, in part at least. . . . [T]he body requires forgetting it' (Serres 1999, 58).

Moreover, the model is that of mixed versus purified bodies, rather than hidden desires versus purified thoughts. But the hierarchy shifts: the ideal is not purification, but mixture. Reality is a mixture and the body must be trained to be affected by it, to experience it. 'No-one has witnessed the great battle of simple entities. We only ever experience mixtures, we encounter only meetings' (Serres 1985, 28). In an interview with Bruno Latour, Serres is explicit about his preference for mixture: 'I'm faithfully pursuing the project of mingling. Notice the title of *Les Cinq sens: philosophie des corps mêlés, vol. 1* [The Five Senses: Philosophy of Mixed Bodies, vol. 1]. You have only to add to all my other books "volume 2", "volume 3", and so on' (Serres and Latour 1992, 165).

A dynamic self is thus a body attentive to and mingled with reality. But what threatens this dynamism? In the case of Serres the threat does not arise from life, but from noise, especially social noise (see Chapter 6). It is not so much a question of what stops thought, but of anaesthetics: What numbs the self? What deafens the body? There is a theme of waking up, but not mentally, but bodily: the body must be woken up in order to feel again.

This brings us to a central element of Serres's proposed model of the self: one can only know something if one makes less noise than the noise produced by the studied object itself. 'The condition for one to be a receiver, a subject, an observer, is that one makes less noise than the noise emitted by the observed object. If one emits more, one erases the object, covers it or hides it' (Serres 1980a, 106). Thus the threat to bodily dynamism comes from ourselves: our collective noise covers and hides the world and makes the body numb.

In the same way power numbs, as illustrated by the tale of Emperor Harlequin in *Le Tiers-Instruit*, who returns from the moon and is asked what marvels he witnessed.[2] Harlequin answers in the negative: nothing new under the sun. The emperor's body is incapable of being affected by the novelties of the moon. The concentration of power has sedated his body. 'Whether royal or imperial, whoever wields power, in fact, never encounters in space anything other than obedience to his power, thus his law: power does not move. When it does, it strides on a red carpet. Thus reason never discovers, beneath its feet, anything but its own rule' (Serres 1991a, xiii). Whereas for Bachelard the autonomy of the human spirit was a prerequisite to become dynamic, it is an obstacle for Serres: an autonomous self is a detached self, and thus numbed.

Similarly, there are a number of revealing passages in *Les Cinq sens* (1985). First you have the anecdote of Serres being stung by a hornet while lecturing:

> If I had been looking at some image, listening to the sound coming from an organ, smelling a garland of flowers, tasting a sugared almond or grasping a pole, the hornet sting would have caused me to cry out. But I was speaking, balanced in a groove or enclosure, protected by a discursive breastplate. You want to anaesthetize a patient completely? Get him to speak with passion and vigour, ask him to talk about himself, and himself alone, of his one desire, demand that he prove something or that he convince his audience. He is intoxicated with sonorous words, the hornet is powerless. Militant egotists, we speak in order to drug ourselves. (Serres 1985, 59)

Secondly, you have the story of the theatre of Epidaurus, where Serres is resting and restoring his body through silence. He refers here to the famous definition of health by the physiologist René Leriche (often used by Canguilhem (e.g. 1989, 91)): Health is life lived in the silence of the organs. However, this silence is disturbed by a group of noisy tourists, who live in their shell of noise: 'They arrived ill, indisposed by the murmuring of their organs, surrounded by their collective noise, and they departed in that clamorous ship without ever really arriving' (Serres 1985, 86). Thirdly, there is the retelling of the death of Socrates: to distract himself from his imminent death, Socrates holds an endless dialogue. 'What do his friends do, hearing him speak while he lies dying? Are they distracting him by making him speak of the soul? Are they anaesthetizing him to pain and fear? Is this dialogue as good as a drug, a narcotic phial' (Serres 1985, 90)?

We see this logic also at work in the retelling of the myth of Argus (Serres 1985, 48). Argus is a giant with a body full of eyes who never sleeps. Recruited by Hera to prevent Zeus from contacting the nymph Io, he is eventually defeated by Hermes. How? By playing the flute and putting Argus to sleep. Besides the theme of numbness, there are other things at work here as well. There is the Foucauldian theme of the panoptic: Argus sees everything. Secondly, there is the theme of mythology: social relations, like myths, are based on a spiral of violence that can only be stopped by finding a scapegoat and sending it away. The social sciences, just as myth, follow a logic of stabbing each other in the back and hoping one is the last. Argus has no back, seemingly thus being the last, yet there is Hermes who still finds a weak spot: numbs his body with noise. This theme of the social sciences, myth and critique, will be taken up in Chapter 6. Finally, there is the ambiguity of Hermes: though typically the positive figure in Serres's philosophy, here he represents a risk: too much noise by our society, in the form of news and advertisements, makes us forgot the object and the world. Serres, therefore, ends with a plea to return to the objects in science through the body, through the skin: after Argus is defeated, Hera pays him respect by putting his eyes on the feathers of the peacock. The eyes of suspicion become the peacock's skin. We return to that theme in Chapter 5.

We thus find a twofold problem with the social: it numbs the body and leads to an endless escalation that swallows all our attention and relations. Both are linked: because we are numbed, we are also numbed for the violence. 'We will stay in the theater while, outside, the nuclear sun ravages the earth' (Serres 1983a, 171). At several places he links this to the cardinal sins, which we encountered in Bachelard (e.g. Serres and Latour 1992, 196). *Variations sur le corps* links the cardinal sins to both the social and the linguistic theme. First of all, 'our culture itself is mistaken for a growing narcosis that is enslaved to its addiction' (Serres 1999, 68). Secondly, they are linked with language and numbers: one collects flattering words and keeps score of sexual conquests, meals or the competition. Serres's contrasts this to the body, the origin of virtue: 'from the body, that is to say from the heart, value comes from courage: from the recognition and rejection of our finitude. The first and only virtue that is valid and from which the others can be deduced, it turns its back on reason as much as invention mocks criticism' (Serres 1999, 71). Courage implies that one ignores the mind, which protests against an action, and acts nonetheless.

This aversion against language and the social suggests that Serres's model proposes to abandon all relations. This can find some support in Serres's recurrent references to themes of loneliness, isolation, moving outside of society and so on that are often used in a positive manner. In Serres's work, we can find positive references to the life of Diogenes and Francis of Assisi, who devote themselves to a life of poverty:

> The bowl can be a gift, the coat can be sold, these things have no more value or still have some, they are exchanged, in gift or damage, graciously or for money. If the bowl is a chalice and the mantle a pallium, if the bowl is the Grail and the cloth is the veil of Tanit, these things, blessed, sacred, are worshiped, we bow down to them. Whether the value of these objects is such or such, we always fight among ourselves to own them, to exchange them, to adore them, to fight again, barely to enjoy them. No things without these collective relationships, no objects without these struggles, these exchanges, this reverence. (Serres 1983a, 122)

What Serres praises in Diogenes and Francis is their detachment: they abandon social relations and escape the spiral of violence. This is also Serres's take on the story of Alexander the Great and Diogenes: Alexander represents power, social relations and violence. He stands between Diogenes and the world, exemplified by the sun. 'Now get out of my sun, Alexander' (Serres 1983a, 128). By stepping outside of society, Diogenes finds the world again, objectivity and science. 'Without Alexander, science takes its sunbath' (Serres 1993b, 121).

But it is mistaken to conclude that Serres's dynamism implies a lonely and isolated individual. Rather it is a dynamism that recreates silence, not for its own sake, but in order for new and other relations to pop up. This is eventually also his take on Diogenes who, similar to Francis, lay at the foundation of new cynic version of Franciscan collectives:

> Pacified, in rags, alone in front of his barrel, showing the zero of usage on the nakedness of his skin, he meditates and asks: can we invent relationships other than struggle, other than exchange or adoration? Can I put my hand or look at something, something that is neither a stake, nor a fetish, nor a commodity? (Serres 1983a, 124)

A new model of purification

We thus find a new model of purification in Serres: one has to free oneself from social relations because they numb our bodies. We have to detach ourselves in order to reattach ourselves: to make room for new ways to be affected by the world. Purification is thus part of the constitution of the self, but it is only a first step.

This model thus differs from that of Bachelard in a number of ways. For one, Serres acknowledges the violence implied by purification, since purification itself is a mythical act (see Chapter 5). 'The more one tries to exclude myth, the more it returns in force, since it is founded on the operation of exclusion' (Serres and Latour 1992, 163). Secondly, the goal is different. For Serres, the starting point is that the world is always full. We find this in Serres's description of agriculture and the countryside in *Détachement* or in the introduction of *Le Tiers-instruit*: Harlequin's coat is full of colours, layer after layer. Because everything is full, one can only survive by creating a new niche or by purification:

> Repetition: all the places are always already taken; everyone comes too late. To seize the held places, certain people then wage war, fight to acquire them, kill to keep them, soon die, must at their death abandon them to their murderers, and the vendetta begins again from generation to generation. (Serres 1993b, 18–19)

Finally, the result is different. One purifies a field, but for it to immediately fill again with new parasites from the outside. This intrusion is what creates novelty. Serres refers again to agriculture: new, cultivated species only arise from an act of eradicating all present lifeforms. Whenever a space is purified something new, a parasite, enters the scene, through a small crack and takes over:

> An emptied square of ground from which the entire plant and forest mantle has disappeared in effect produces an abrupt divergence from equilibrium in the life of the flora interlaced around abouts. Through this tear or rift passes the vertical proliferation of a given single species, sown there by the winds of chance. Since no obstacle stands before it, it crosses the percolation threshold; its flow begins. (Serres 1993b, xlii)

The same model applies to the origin of geometry: 'white, this *page* designates in our languages, from the same word as *pagus*, the space, of wax or papyrus, where writing was born in the same way' (Serres 1993b, xliv). Born from a blank page, purified from social preoccupations, something new can come into being.

We encounter here the theme of the third: we have to exclude a third, in order for something new, fragile and local to come into being. But this exclusion must be acknowledged and can never be permanent and complete, since soon enough new parasites will enter the scene, raising the question of purification anew. The central question becomes the examination of relations: which relations should we foster and which should we reject? In subsequent chapters we will see that this ethics of relations is central in Serres's analysis of the quasi-object: entities that are neither isolated nor passive, but instead are shaped by the numerous relations to other entities surrounding them. But how to cultivate quasi-objects? How to prevent them from becoming parasites? As we shall see, this is how Serres will approach the issue of our collective destruction: 'We do not yet have an adequate idea of what the deluge of objects manufactured since the industrial revolution by science, technology, laboratories, and factories implies for our relations-and now for those universal relations brought about by our global enterprises' (Serres and Latour 1992, 202). Finding a way to cultivate the right relations will become the aim.

The self-forming activities of the scientific body

Once again the question remains what the self-forming activity is, the means to change the self. In contrast to Bachelard, Serres's work seems to offer no concrete technologies of the self. But another road is to draw again inspiration from Foucault: one specific kind of technology of the self the latter describes is that of reading and writing itself. Within Greco-Roman philosophy texts played a different role than what we now associate with them: 'the object or end of philosophical reading is not to learn an author's work, and its function is not even to go more deeply into the work's doctrine. Reading basically involves – at any rate, its principal objective is – providing an opportunity for meditation' (Foucault 2005, 356).

Meditation must be understood in a particular sense. It does not refer to simply reflecting on a certain topic, but to a different practice. 'The *meditatio* involves, rather, appropriating [a thought] and being so profoundly convinced of it that we both believe it to be true and can also repeat it constantly and immediately whenever the need or opportunity to do so arises' (Foucault 2005, 357). Secondly, meditation is an experiment of identification: the self does not simply think about something, such as its own death, but practices the thing it is thinking about. It is 'not a game the subject plays with his own thought or thoughts, but a game that thought performs on the subject himself. It is becoming, through thought, the person who is dying or whose death is imminent' (Foucault 2005, 357–8).

Can we similarly read Serres's own texts as meditations, whose goal is to introduce a transformation in the reader? This might explain the particularity of his texts, which indeed are never systematic elaborations on certain topics, but explorations of certain themes that aim to provoke a response in the reader. Moreover, I want to suggest that the meditation is not so much, as with Foucault, an attempt to let the reader take the position of another subject, but rather that of another object, a quasi-object.

A book such as *Parasite* focuses on the phenomenon of the parasite, not so much in order to give you a theory about it, but to let you 'practice' the phenomenon: reproduce the parasitic logic in order to make the self capable to bring it up, detect it, whenever it presents itself to the self. Similarly, *Les Cinq sens* is not so much a theory of the five senses, but encourages the self to experience the world anew through all of its senses. The goal is to transform oneself, to teach the body to be affected by the world in new ways, by meditating on the position of the phenomenon, the quasi-object.

Conclusion

In the previous three chapters we have seen how an understanding of Serres's work implies an examination of the broader tradition of French historical epistemology. We thus encountered a number of shared themes (such as surrationalism or phenomenotechnique) and a number of differences (such as the polemics about the Althusserian legacy and the ethos of purification). But these chapters also highlighted how Serres's work can help us understand the tradition of historical epistemology, as is shown in some detail in this chapter.

In the next five chapters, I will turn to Serres's own alternative and how his work has initiated a number of new approaches and themes in French philosophy of science. It thus aims to answer the question: What can be done with the alternative normative model of the scientific self? We will see how Serres develops another ontology (Chapter 4); a different conception of time and modernity (Chapter 5); an anthropology of science disclosing the hidden violence in science (Chapter 6); and an alternative ecology of relations, that aims to overcome this violence, either through religious themes (Chapter 7) or via an ecology of quasi-objects (Chapter 8).

4

French object-oriented philosophy in the 1970s: Serres, Dagognet and Latour

Introduction

In the previous chapters we have mainly focused on how Michel Serres's philosophy is related to his predecessors. In this chapter we are turning our gaze forward to his own alternative philosophy and the impact it has had on other scholars. This impact is probably best seen in the work of Bruno Latour, though it is not often recognized. As a consequence there are many misunderstandings of their work. The most striking paradox is how Latour is situated on the realism vs. social constructionism landscape, where he has been attacked from both sides. This has been noted by some commentators, such as Graham Harman:

> [Latour is] attacked simultaneously for opposite reasons. For mainstream defenders of science, he is just another soft French relativist who denies the reality of the external world. But for disciples of Bloor and Bourdieu, his commerce with non-humans makes him a sellout to fossilized classical realism. (Harman 2009, 5)

Although Latour's thought is typically seen as a radical form of social constructionism (Hacking 1999; Kukla 2000), he is simultaneously being criticized by sociologists for relapsing into a naïve form of realism (Collins and Yearley 1992). Similarly, Yves Gingras has claimed that Latour's position falls back on a traditional notion of nature. 'I have nothing against this form of realism as long as it is acknowledged as such and not presented under the guise of actor-network language' (Gingras 1995, 131).

In a similar vein, Latour's work has been perceived as one of the so-called 'new realists, including [Quentin] Meillassoux, [Levi] Bryant, and Latour' (Gratton 2014, 120). Or, Harman claims that 'the most important recent French thinker for [him] is Latour' (Harman 2018a, 198). This paradox, however, reproduces itself within speculative realism, since some speculative realists accuse Latour of being an anti-realist, reducing things to descriptions, concepts and language (e.g. Ferraris 2014, 32). This criticism is echoed by other authors, surprisingly even by Pierre Bourdieu himself. According to him, Latour starts from the 'semiological model' of A. J. Greimas (1976), leading to the premise that '[b]ecause scientific facts are constructed, communicated and evaluated in the form of written statements, scientific work is largely a literary and interpretative activity' (Bourdieu 2004, 27). For Bourdieu, this assumption leads to a form of linguistic idealism or what he calls *textism,* 'which constitutes social reality as text' (Bourdieu 2004, 28). A similar critique is present in Baird (2004). Although far more sympathetic to Latour and Woolgar (1979), Davis Baird argues that:

> The picture Latour and Woolgar present of science is thoroughly literary. 'Nature', with the help of 'inscription devices' (i.e., instruments), produces literary outputs for scientists; scientists use these outputs, plus other literary resources (mail, telephone, preprints, etc.), to produce their own literary outputs. (Baird 2004, 7)

For Baird, such a 'picture of the function of a laboratory is a travesty.... Laboratories do not simply produce words'. Latour and Woolgar are thus 'continuing a long tradition of text bias, they misdescribe the telos of science and technology exclusively in literary terms' (Baird 2004, 7).

We are thus faced with a strange paradox in which Latour is being criticized both for being a realist and for being a social constructionist. The goal of this chapter is to solve this paradox by tracing back some of Latour's roots in the work of Michel Serres and other French philosophers. Too often his work is seen as a mere exemplar of Anglo-American social constructionism, linked to the Strong Program in British sociology (Bloor 1976; Barnes, Bloor and Henry 1996). Despite the fact that Latour occasionally situates himself in this tradition (Latour 1999c, 114), by the time he came into contact with this work, around 1976, he was already in his late twenties and had a whole academic training, including a PhD, behind him. In a recent interview, referring to these early days, Latour confessed: 'I discovered then, through Steve Woolgar, the existence of the English school of the Edinburgh school, and then I started reading' (Mazanderan and Latour 2018, 285). In the same series of interviews, Woolgar himself commented on the genesis of *Laboratory Life*:

> Bruno's English was nothing like as good as it is now. So the book was very much his fieldwork but my writing. That was the basis of our collaboration. Also, he didn't know much of the literature in the sociology of science at the time. I think he knew Merton and a few things like that, but he was really unaware of

the sociology of science, so that was what I brought to the writing of the book. (Mikami and Woolgar 2018, 306)

Nonetheless, it would be wrong to claim that Latour merely brought in the ethnographical data while Woolgar was responsible for all the theory. Already in the early books of Latour (Latour and Woolgar 1979; Latour 1984, 1987) a strong French philosophical tradition is present (Mialet 2012; Schmidgen 2014). In the final chapter of *Laboratory Life,* for example, Latour and Woolgar aim to 'identify six main concepts' (Latour and Woolgar 1979, 236). It is remarkable how French all these concepts are: *Agonistic* is derived from the work of Jean-François Lyotard (see next chapter), *reification* from Jean-Paul Sartre (1943), *credibility* from Bourdieu (1975), *circumstance* from Michel Serres (1974a) and *noise* from Léon Brillouin (1956). Even *construction,* which they borrow from Karin Knorr-Cetina (Knorr 1977), is simultaneously linked to Gaston Bachelard (see Chapter 2).

Each of these cases of conceptual borrowing would deserve a treatment of its own, an ambition that exceeds this chapter by far. Here, I will limit myself to two authors: François Dagognet and Michel Serres. There are several reasons for this. First of all, as I will argue, it is from them that Latour derived a number of central concepts, such as *inscription* and *translation*, crucial to understand how Latour can seemingly be both a realist *and* reduce everything to texts. Secondly, highlighting these relations will show how Latour, though often labelled as a postmodernist, can be linked with speculative realism as well. Finally, it will introduce the work of Dagognet, who has so far not received much attention outside of France, and further help to elaborate that of Serres. Methodologically, I will limit myself to the early Latour, focusing on his work before *Nous n'avons jamais été modernes* (1991). This is because Dagognet's and Serres's influence is particularly present in this early work.

This chapter is structured as follows: First, I will focus on Dagognet and the notion of inscription, which is central to Latour. I will show how this concept leads to a position that is paradoxically both focused on texts and on material objects. Next, I will argue that a similar importance must be attributed to Serres, who in his own work develops a novel ontology, centring on the concepts of circumstance, translation and the parasite. Indeed, what would later become actor-network theory (ANT) was originally called the sociology of translation, mainly due to Michel Callon (1975, 1984). But similar to the concept of inscription, 'translation' suggests a philosophy focused on language. However, as we will see, Serres himself is a philosopher who is calling for a return to the object. Therefore, as I will argue, what initially looks like a paradox or a weakness – namely that it is unclear whether Latour speaks about things or about words – is in fact a strength of this alternative perspective, as developed by Dagognet and Serres.

François Dagognet and inscriptions

In the early work of Latour the concepts of 'inscription' and 'inscription device' play a central role. The idea of an inscription device is introduced in *Laboratory Life* (1979) and refers to the apparatuses that 'transform pieces of matter into written documents. More exactly, an inscription device is any item of apparatus or particular configuration of such items which can transform a material substance into a figure or diagram which is directly usable by one of the members of the office space' (Latour and Woolgar 1979, 51). They derive this term from the concept of inscription, through which 'the laboratory began to take on the appearance of a system of literary inscription' (Latour and Woolgar 1979, 52).

In a footnote, Latour and Woolgar (1979, 88n2) state that the concept of inscription is inspired by Jacques Derrida's *De la grammatologie* (1967). Derrida is focused on deconstructing what he calls Logocentrism or Phonocentrism: the belief that speech is superior to the written word and that the latter is a mere copying or even a shadow of the former. Derrida links this to his broader project of criticizing a metaphysics of presence, where what is present, visible and in the open, is seen as superior to that which is only accessed indirectly, momentarily or remains (partly) hidden. To counter this, Derrida will speak of inscriptions or 'writing' in general, as a new general category of which the spoken word is seen as one instance: 'the concept of writing . . . is beginning to go beyond the extension of language. In all senses of the word, writing thus *comprehends* language' (Derrida 1967, 7). The aim is to stress a number of characteristics of any form of 'writing' that are forgotten in a phonocentric framework, such as materiality, impurity, variability and lack of control. In Derrida's own philosophy nothing is purely present, but always carries a trace of references to other significations and meanings. Such a perspective is often read as a reduction of everything to language, illustrated by the infamous quote: '*There is nothing outside of the text*' (Derrida 1967, 159).

A similar reading could be given of *Laboratory Life,* which also stresses 'the omnipresence of literature' (Latour and Woolgar 1979, 53). The book claims that one has to understand 'the laboratory in terms of a tribe of readers and writers who spend two-thirds of their time working with large inscription devices' (Latour and Woolgar 1979, 69) and that although scientists themselves 'claimed merely to be scientists discovering facts' the visiting anthropologist 'doggedly argued that they were writers and readers in the business of being convinced and convincing others' (Latour and Woolgar 1979, 88). This is especially the case if one focuses on how Latour and Woolgar, inspired by Greimas (1976), study the shifts in modalities of sentences in scientific texts: 'A laboratory is constantly performing operations on statements; adding modalities, citing, enhancing, diminishing, borrowing, and proposing new combinations' (Latour and Woolgar 1979, 86–7). These statements themselves seem to come close to some form of linguistic idealism where making a statement makes something real.

However, another reading of their work is possible, which starts from François Dagognet. Latour and Woolgar refer to Dagognet to stress that the notion of inscription 'designates an operation more basic than writing (Dagognet 1973). It is used here to summarize all traces, spots, points, histograms, recorded numbers, spectra, peaks, and so on' (Latour and Woolgar 1979, 88n2). Latour himself would later say that he 'was helped not only by Derrida but also by François Dagognet, whose little book *Écriture et iconographie* (1973) put me on the right track: I followed it like a hunting dog, nostrils flaring' (Latour 2013b, 290–1). According to Latour, it was 'one of the few books I read carefully when I wrote *Laboratory Life*' (Latour 2006, 119). Moreover, Latour expresses his enthusiasm over Dagognet's work at several instances, claiming that 'François Dagognet's work represents the best of the Bachelardian tradition' (Bowker and Latour 1987, 734–5). Latour situates Dagognet in a French tradition of philosophy of technology, 'a very rich tradition from Diderot, Lafitte, Bergson, André Leroi-Gourhan all the way to François Dagognet, a lesser known but quite interesting figure' (Latour 2007, 126).

Similar to Serres, François Dagognet (1924–2015) was a student of Georges Canguilhem, trained in philosophy and medicine. His double dissertation dealt with pharmacology (1964) and with Louis Pasteur (1967). Just like Serres, Dagognet soon turned his eye to new developments in chemistry and biology and the increasing influence of information theory and data science (Dagognet 1973, 1979). A prolific writer, he would write books on topics as diverse as art, the body, industry and science. Dagognet always focused on the analysis of concrete objects and phenomena, as he explicitly puts it in his 'materiology' in the 1980s (Dagognet 1989). Dagognet was one of the few early allies of Latour in France. He knew Latour and presided over Latour's jury of his *agrégation de philosophie* in 1972 (Mialet 2012, 457).

At first sight, one could find in Dagognet a similar linguistic idealism. Speaking about contemporary chemistry, for instance, he argued that 'all reality becomes speech. The entire world is word' (Dagognet 1969, 158). But this is again a very one-sided reading. In the work of Dagognet, the concept of inscription is less linked with language and culture than with developments in the natural sciences. There are several ways to understand this. A first and weakest sense is to argue that Dagognet claimed that the natural sciences are transforming in a form of linguistics or philology, since we are witnessing a number of revolutions through which the role of computers, data and text increases:

> Revolution: the word and the figure – the audio-visual – define the new symbolic or expressive chemistry. The world is translated, transported. It passes from the reality where it was spread to a discourse that records it, letters that pick it up, a tableau that reflects it. An immense translation. (Dagognet 1969, 213–14)

Dagognet has been criticized for overemphasizing the novelty of recent developments in chemistry. Ursula Klein (2003), for example, argued that Berzelian formulas already played a similar productive role in the nineteenth century. '*Pace* Dagognet, most chemists of the 1960s did not spend their time assembling "books of books" with the help of computers, because specialists had already been doing so for decades before they first brought computers to bear upon this task' (Hepler-Smith 2018, 17).

However, a second, stronger claim is equally possible: what Dagognet was after is not specific to recent scientific disciplines, but present in all sciences. More specifically, the focus on inscription plays a role in two topics of concern that Dagognet focused upon, and that will reoccur in the work of Latour: the material embodiments of these inscriptions and their internal and paradoxical productivity.

The materiality of inscriptions

One of the central insights from Dagognet's work is that in scientific practices a specific form of materiality plays a crucial role. Latour acknowledges the importance of Dagognet for this insight. What Dagognet teaches us is that '[t]he mistake before was to oppose heavy matter (or "large-scale" infrastructures like in the first 'materialist' studies of science) to spiritual, cognitive or thinking processes instead of focusing on the most ubiquitous and lightest of all materials:

the written one' (Latour 1983, 162). Similarly, in another article, Latour claims: 'As Dagognet has shown in two excellent books, no scientific discipline exists without first inventing a visual and written language which allows it to break with its confusing past (1969, 1973)' (Latour 1986, 14). Dagognet was indeed fascinated by the material embodiment of the inscription:

> For me, it is incomparably important that *on* which and that *by which* one writes. One is inclined to underestimate it, if not to forget it. Nothing but the meaning would count, but the meaning does not really break free from that which conditions it, or in any case, carries it. In short, the substrate deserves our attention: it ends up deciding the rest. (Dagognet 1979, 70)

From this perspective, science is a 'profoundly modifying and illuminating codification' in which 'a flower becomes a diagram, an animal a drawing' (Dagognet 1973, 47). This moment is not restricted to new disciplines or developments, but is present in all sciences:

> There is no discipline, indeed, which does not benefit from iconicity: from physics, kinematics, to geology, technology or even physiology. Everywhere drawings, trajectories, contour lines, maps, in short, structural and geometric figures are required. The mistake would be to regard them as mere didactic aids, as convenient illustrations, while in fact they constitute a privileged heuristic instrument: not an ornament, a simplification, or a pedagogical means to facilitate transmission, but a true new form of writing, capable, alone, of transforming the universe and inventing it. (Dagognet 1973, 86)

In *Écriture et iconographie* this led to a reconceptualization of science, focused upon this neglected dimension of scientific practices, but 'what I believe to be one of its founding moments, the one by which it makes us truly masters of the universe that surrounds and overflows us: the geometric and abrivative iconicity, a certain writing that transposes the world, projects it and renews it' (Dagognet 1973, 7). From this perspective, '[t]o know is not to measure or to analyse – these are secondary tasks – but first of all to write the phenomena in a spatial, universal language of traces, trajectories and curves. Physics is the first science of "translation-inscription", a kind of super-philology' (Dagognet 1973, 110).

We find a similar argument in other books from this period, such as *Mémoire pour l'avenir* (1979). Here Dagognet reflects on the newly arising discipline of data science, but again to highlight a lesson applicable to all sciences. All scientific disciplines incorporate, according to Dagognet, three moments: the recording of information, the transportation of information (memory) and the modification of this information in order to create something new. Scientific practice thus deals with information in a threefold way: 'collection, conservation and exploitation' (Dagognet 1979, 7). Dagognet is mainly fascinated by science's power of memorization and which allows it to compare, remember and explore. 'Thus, humanity's power consists in substituting for an enslaving multiplicity a representation that can integrate, contract and make visible. By what ruse it is able to do so, that is our question' (Dagognet 1979, 20).

The same question also becomes central in the work of Latour. Within the laboratory something remarkable happens: chemical liquids, animal samples and instrumental manipulations are converted into a specific material form: that of a graph or a text. For Latour, to focus on texts *is* to focus on the material setting. 'Epistemologists had chosen the wrong objects, they looked for mental aptitudes and ignored the material local setting, that is, laboratories' (Latour 1983, 160). Even when Latour and Woolgar were speaking about 'literature' in *Laboratory Life* they understood it materially:

> 'Literature' refers both to the central importance accorded a variety of documents and to the use of equipment to produce inscriptions which are taken to be about a substance, and which are themselves used in the further generation of articles and papers. In order to explicate the notion of literary inscription as applied to apparatus, we shall provide an inventory of the material setting of the laboratory. (Latour and Woolgar 1979, 63)

One should thus focus on the inscription devices, the material settings that allow the world to be translated into visible and material traces which scientists can decipher. 'The invisible becomes visible and the "thing" becomes a written trace they can read at will as if it were a text' (Latour 1983, 163). Inscriptions are thus not about ignoring the world in favour of texts but highlight the moment where world and text meet. 'Through the laboratory, the text and the spectacle of the world end up having the same character' (Latour 1986, 22). What first seemed as a confusion in Latour between realism and linguistic idealism thus becomes a productive combination which allows for a new perspective on why scientific practices are so successful:

> The two-dimensional character of inscriptions allows them to merge *with geometry*. . . . Better still, because of this optical consistency, everything, no matter where it comes from, can be converted into diagrams and numbers, and combinations of numbers and tables can be used which are still easier to handle than words or silhouettes (Dagognet 1973). You cannot measure the sun, but you can measure a photograph of the sun with a ruler. (Latour 1986, 22)

The paradoxical productivity of inscriptions

A second crucial lesson that Latour draws from Dagognet is what one could call the paradoxical productivity of inscriptions. In a later text, Dagognet summarized this paradox as follows:

> We believe even to have shown that there is more in the translation or the schematized portrait than in the model; the theorist, in the presence of a graph, draws more information than if he scrutinized the thing; Why? (Dagognet 1998, 19)

The paradox of how a representation, summary or scheme can show *more* by showing *less* is an often-repeated theme in Dagognet's work. Dagognet is fascinated by 'the theory according to which "the image" of a thing does not reiterate, but renews or modifies it' (Dagognet 1975, 8), 'how statements can become new without really changing' (Dagognet 1979, 8) and how 'to change the representation of reality, is to transpose and transform reality' (Dagognet 1973, 87). Dagognet saw it as a widespread methodological error to assume that

> one reads or finds on a map only what one has previously written on it! A simple decoding. In reality, at a certain moment, the route is metamorphosed: instead of gathering only indices, it turns into an indicator. It leads to fruitful correlations, elicits extrapolations, leads by itself to complete the profiles, to straighten the visibly lowered or dislocated slopes, in short to re-write the various parallel lines that are too often displaced or interrupted. (Dagognet 1973, 113)

Again, chemistry is the most intuitive case, where 'the *reproduction* was an authentic *production:* it gave the key to knowledge. The work of Lavoisier and even more Mendeleev have verified this. For all these reasons, the world did not so much enter the Periodic Table as came forth from it' (Dagognet 1969, 217).

A convincing explanation of this paradox is not really present in Dagognet. At the same time, this paradox is not particular to Dagognet. We already saw similar themes at work in Gaston Bachelard's philosophy (see Chapter 2). Yet, this constructive aspect of scientific practices is often not recognized. Dagognet acknowledged this neglect and tried to explain it by again referring to Derrida and his argument that we tend to desire immediacy and directness, while ignoring and neglecting mediation (Dagognet 1973, 31).

Latour takes a different, novel path, one that will become central in his later writings. Already in *Laboratory Life* he and Woolgar reflect on how an 'important feature of the use of inscription devices in the laboratory is that once the end product, an inscription, is available, all the intermediary steps which made its production possible are forgotten' (Latour and Woolgar 1979, 63). They link this with the resistance towards sociological explanations of scientific facts: only if this constructive aspect is recognized could such a sociology of scientific knowledge make sense. In their book they thus not only want to argue that facts are constructed but 'also wish to show that the process of construction involves the use of certain devices whereby all traces of production are made extremely difficult to detect' (Latour and Woolgar 1979, 176). This is one of the reasons why they delve into discussions about shifting modalities of statements that hide the constructive elements. Although still lacking a name here, Latour will later capture this aspect under the banner of 'purification', which will become a central characteristic of Latour's definition of modernity (see Chapters 2, 5 and 8).

Michel Serres and translations

There is a second crucial concept in Latour's work derived from French philosophy that plays a fundamental role, namely that of *translation*. Similar to the concept of inscription, translation seems to suggest a language-focused philosophy. However, as we will see, the concept is derived from Michel Serres who, similar to Dagognet, aims to develop a position which combines a focus on language *and* on things. Most commentators overlook this link to Serres and understand translation as a purely linguistic phenomenon (e.g. Gratton 2014, 90). Even Graham Harman, who discusses the notion of translation for several pages, fails to mention Serres (Harman 2018b, 57–8). Nonetheless, Latour is very explicit about Serres's influence on his work (Latour 1987b), claiming that within France 'Michel Serres is the only one to contribute to a rethinking of science, of discourse and, in many ways, of society' (Bowker and Latour 1987, 730). Serres remains a constant reference, including in Latour's recent books on Gaia (see Chapter 8).

Some commentators have noticed this connection. In an early review of Latour's ANT, Chateauraynaud, for instance, notes how this 'new sociology has been constituted by a massive importation of formulas elaborated on by Michel Serres' (Chateauraynaud 1991, 463). More recently, Henning Schmidgen claims that 'Latour has repeatedly cited Michel Serres to elucidate the central motifs of his work, for example, the subject of translation or the alleged continuity between early and late modernity' (Schmidgen 2014, 5).

This influence is already present in Latour's early publications. Schmidgen highlights how 'the end of *Laboratory Life* anticipates Latour's exploration of

Michel Serres's quasi-cybernetical philosophy' (Schmidgen 2014, 44). But it is even more explicit in another early article by Latour (1981). In this article Latour complains that sociology lacks a theoretical framework to acknowledge that disorder is common and omnipresent in scientific practices:

> The only attempt has not been made by sociologists of science, but by isolated scientists dealing with information, or with turbulent phenomena. The works of Brillouin and the recent book of Prigogine and Stengers together with the philosophy developed in France by Michel Serres, convinced me that a new framework is already at hand to understand and rephrase our observations on the way science is made. (Latour 1981, 70)

Latour thus explicitly states that his alternative approach to understanding scientific practices is inspired by Serres and related thinkers. He draws on an informational ontology developed by Serres and inspired by information theory and molecular biology. As we will see, the notion of translation is not yet present in Latour's early texts, which are rather focused on other concepts such as circumstance and noise. Only in later texts, and due to the influence of Michel Callon and his project of a 'sociology of translation' did that notion became more prominent.

Information theory and molecular biology

As we have seen, Michel Serres was trained in mathematics and initially aimed to write his doctoral dissertation on algebraic topology. However, he 'discovered' Leibniz and changed topics (Serres 2014, 44–5). In Leibniz he discovered a thinker of networks and communication, later claiming that his 'whole thesis on Leibniz was founded on the idea of the network' (Serres 2014, 125).

In the 1970s Serres would publish his *Hèrmes* series, a number of books with commentaries and interpretations of literary texts and scientific developments, linked together by the theme of communication and information. In these books Serres, based on developments in mathematics, information theory and biology, aims to develop what one could call a new ontology and to subsequently apply it to the history of philosophy, literature and science. Serres claims to discover a 'new Leibnizianism', starting from the idea that 'contemporary knowledge, in its totality, *is* a theory of communication' (Serres 1974a, 41).

One of his main inspirations is the French physicist Léon Brillouin (1889–1969). In his later life Brillouin wrote a number of books on the new science of information theory (Brillouin 1956). Based on his reading of these books, Serres claimed that

> the theory of information has constituted a philosophy of physics, intrinsic to the discipline itself. It is remarkable, for instance, that Brillouin has chosen as the title of his last work: *Science and Information Theory*. One finds in this work, indeed, a complete, descriptive, quantified, normative and founding epistemology, expressed in the language of physics itself, of the notion and practice of experimentation, scientific laws, precision and approximate knowledge, the limits of knowledge (what can I know?), in short all the classical topics; and all the 'modern' ones: a theory of codes, language, writing and translation. Philosophers need neither look for nor write a handbook of the epistemology of experimental knowledge: it is found here. (Serres 1974a, 45)

Let us have a brief look at the work of Brillouin himself. He defines information as 'a function of the ratio of the number of possible answers before and after' (Brillouin 1956, x). An increase in information entails a decrease in possible solutions.[1] If I am, for instance, wondering what the capital of France is, receiving the information that it is not Marseille or Lille, or that it starts with a P, narrows down the possible answers. But Brillouin mainly defines information in this manner to accomplish something else: to link the concept of information with that of entropy. In doing so he thus aims for a 'generalization of the second principle of thermodynamics' (Brillouin 1956, xii):

> Every physical system is incompletely defined. We only know the values of some macroscopic variables and we are unable to specify the exact positions and velocities of all the molecules contained in a system. We have only scanty, partial information on the system, and most of the information on the detailed structure is missing. Entropy measures the lack of information; it gives us the total amount of missing information on the ultramicroscopic structure of the system. (Brillouin 1956, xii)

Whereas entropy is defined as the lack of information, information itself can be defined as 'negative entropy', a quantity for which the author coined the word negentropy' (Brillouin 1956, vii). From this perspective, information theory offers a solution to an infamous problem in thermodynamics, namely Maxwell's demon. In 1867, J. C. Maxwell came up with this thought experiment to show how the second law of thermodynamics might be violated. In this thought experiment, a demon controls a trapdoor between two chambers filled with gas. The demon can open the door when fast particles are approaching while keeping it closed for slower molecules. The demon therefore seems capable to decrease entropy in one of the rooms and thus violates the second law.

According to Brillouin, information theory 'enables one to solve the problem of Maxwell's demon and to show a very direct connection between information and entropy' (Brillouin 1956, vii). The reason for this is that information theory shows that the actions of the demon itself would come with an informational cost: either the demon simply translates energy into information, thus keeping the entropy constant, or, more likely, increases entropy through the act of measuring the heat of the atoms and controlling the trap door.

This alternative take on this thought experiment offers Brillouin a way to analyse the practice of scientific experimentation itself: doing an experiment is itself a matter of creating negentropy by producing entropy outside of the experimental system. It is in this sense that Serres's remark, that we find in Brillouin's work an epistemology internal to the theory of physics itself, must be understood. What he takes from Brillouin is 'the negentropy principle of information, which says that any information resulting from a physical observation must be paid for by an increase in entropy in the laboratory' (Brillouin 1956, 229). Brillouin spends several chapters showing how this principle is at play in scientific experimentation, ranging from the measurement of length, time intervals or counting to the activity of reading and writing: 'We need a source of light to read a book. The phonograph does not work without a motor to turn the record. . . . In all of these examples, the amount of negentropy taken from the external source can be demonstrated to be larger than the information obtained' (Brillouin 1956, 261).

These insights are mobilized by Serres and Latour to develop their philosophies of science. In doing so, they divert from Brillouin's own intentions. For Brillouin, there was an important restriction, namely that his information theory completely ignores the human value of information and thus the question of meaning. He recognizes that his theory is unable to make a difference between a sentence of 100 randomly selected letters and a 100-letter sentence from Shakespeare or a theorem of Einstein. 'In other words, we define "information" as distinct from "knowledge", for which we have no numerical measure. We make no distinction between useful and useless information, and we choose to ignore completely the value of the information' (Brillouin 1956, 9). Whereas Brillouin recognizes the exclusion of the human subject and meaning as a weakness of his theory, this will be translated as a strength in the cases of Serres and Latour. They will argue that it is showing that we can develop a convincing theory of science and the world that does not require this classical concept of the subject. It gives us evidence that we, in fact, do not need these concepts and that an anti-humanism makes perfect sense.

A second resource on which Serres draws is that of recent developments in molecular biology. As we saw (see Chapter 2), molecular biology was picked up by several philosophers, such as Michel Foucault (Talcott 2014; Erdur 2018), Dagognet and Derrida,[2] who argued that science is a matter of not only inscriptions but nature itself: the whole world is information. The clearest case however is Serres, who wrote extensive reviews of both François Jacob's *La logique du vivant* (1970) and Jacques Monod's *Le hasard et la nécessité* (1970). It is in the review of Monod's book that the earlier remarks on Brillouin are written. In the other review, Serres notes how contemporary biology is a 'return' to earlier times, where 'science was nothing but the deciphering of a code' (Serres 1974a, 21). Molecular biology has thus 'discovered this, in its region, like logic, mathematics, physics and chemistry had discovered it in their regions, each in turn. It designates, like the other sciences, a global philosophy of marked elements. It is enough to listen to how it phrases itself: it speaks of a *universality of the code*. This is an expression of Leibniz' (Serres 1974a, 21).

Towards a new informational ontology

In his *Hèrmes* series, Serres generalizes these insights into a new ontology, in which all objects exchange information:

> A new, painful blow to human narcissism: all things in the world exchange information and store it. All bodies are engraved, such as Rosetta Stones or graphic sheets. And that is their specificity, e.g. the shape of the crystal or the elementary composition. And some adapt to each other, and how can we describe this otherwise but to say that they read each other? So the transformations specific to matter are done by writing and reading: to turn upside down what seems to be anthropomorphism, the consequence is rather that we act like them. (Serres 1977a, 217)

From this perspective, a different conception of the living world follows. For Serres, 'Life is communication' (Serres 1974a, 51) and an 'organism is not so much a set of generic elements, of independent particles, as a set of relations, arrangements, combinations' (Serres 1974a, 24). Even the subject is reconceptualized, being just one of the information exchanging entities among others: 'Something which reads and is read like any other, which is written and sometimes writes. Thus, by inversion or by reflection, the object can be called me, and speaks, like me. Let its voice come, its clamour, its autonomous information' (Serres 1977a, 159).

For Serres, the whole world can be understood through the lens of information theory, starting from the idea that communication consists not so much in establishing a connection between sender and receiver but rather in excluding all the background noise and leaving only the meaningful relations intact and stable, even if only temporarily. As Serres already invokes in his first *Hèrmes* book: 'To hold a dialogue is to suppose a third man and to seek to exclude him; a successful communication is the exclusion of the third man' (Serres 1969, 41).

From this angle, the universe consists in a number of ontological layers, atoms, cells, bodies, consciousnesses, each exchanging and constituted by information. Human language is just the end of this long chain. But more importantly, 'I do not know who is the first transmitter, at the other end. It is indefinitely a black box. A box of boxes, and so on.... All I know, but I'm sure of that, is that they are all structured by the pairs of information-background noise, chance-program, entropy-negentropy' (Serres 1977a, 270). Every meaningful layer thus comes forward out of and stabilizes itself from the 'noise' of the previous layer: cells as temporary stabilized atoms, humanity as a temporary stabilized species, knowledge as a temporary stabilized system of brain cells or social relations:

> Nothing distinguishes me ontologically from a crystal, a plant, an animal and the order of the world: we drift together towards the background noise of the universe, and our various complexions of this system go up the entropic river towards the solar source, itself derived from this river. Knowledge is at most only a reversal of this drift, this strange exchange of time, always paid for by drifting, but it is the complexity itself, what one also called being. (Serres 1977a, 271–2)

This view has consequences for our conception of physics and science. We get a reversal, where disorder is standard and order is the rare and exceptional case. 'Order is not the norm, it is the exception. The physicist has become anarchist, the background noise is his domain, where the music is rare. Every object must be called a miracle' (Serres 1974a, 49).

Just as for Brillouin, this leads to a different conception of what scientific practice is about. First of all, science is a mere continuation of the process, a new layer of information, which aims to 'slowly acquire the innumerable codes by which objects behave, it is an attempt to speak the multiplicity of objective languages' (Serres 1977a, 217). In that sense, 'the real produces the conditions and means of self-knowledge' (Serres 1977a, 271). But secondly, the act of knowing is also no longer seen as a passive registration, but as an active intervention, which creates new information and relations, but at the cost of transforming parts of the world and introducing new forms of entropy. This was the lesson from Brillouin: 'All knowledge has a price, a cost, indexed, on the label. Whatever it may be, it is always evaluable and can be recorded in an overall balance sheet of income and expenses. The theory of knowledge is the tableau of this balance sheet' (Serres 1977a, 35).

In his own work, Latour mobilizes this framework to develop his sociological approach of scientific practices. Whereas the elements of negotiations, uncertainty, struggle and failure are dismissed in traditional accounts of science as mere 'disorder' that distracts from the underlying order, Latour uses Serres's view to turn the tables: 'if you eliminate the opportunism, the context, the fiction building, the agitation, the reconstruction, the rationalization you get *nothing* at all; if you introduce them you understand how the scientific facts, discoveries and theories emerge and are maintained' (Latour 1981, 70). It is out of this disorder that the order of a scientific fact can come forward. This is clear in *Laboratory Life*, especially in how they phrase their own goal, namely to understand how 'scientific order is constructed out of chaos' (Latour and Woolgar 1979, 33), and map the 'disordered array of observations with which scientists struggle to produce order' (Latour and Woolgar 1979, 36). Indeed, they draw explicitly on the work of 'Michel Serres, who, in turn, ha[s] been greatly influenced by authors such as Brillouin and Boltzmann and by new developments in biology' (Latour and Woolgar 1979, 251).

One way in which they mobilize information theory is through the concept of 'noise (or more exactly, the ratio of signal to noise)' (Latour and Woolgar 1979, 239). As we have seen, according to Brillouin an increase in information implies a decrease in possible solutions or interpretation. Latour and Woolgar apply this to the question of how scientific hypotheses compete with one another. 'If a large number can easily be thought of, the original statement will be taken as meaningless and hardly distinguishable from others. If the others seem much less likely than the original statement, the latter will stand out and be taken as a meaningful contribution' (Latour and Woolgar 1979, 240–1). The difference between these statements is, however, not something that occurs naturally, but implies a process of intervention, which 'transforms any set of equally probable statements into a set of *unequally* probable statements' (Latour and Woolgar 1979, 241). Thus by creating information, and increasing the negentropy, one hypothesis can become more plausible than the other.

Latour and Woolgar want to focus on 'the imposition of various frameworks by which the extent of background noise can be reduced and against which an apparently coherent signal can be presented' (Latour and Woolgar 1979, 36–7). This is what they have in mind when they speak of a 'social' construction of scientific facts. Social factors are thus invoked to narrow down the possible answers, thus providing information in the sense that Brillouin understood it: 'In principle, the number of alternative readings of [a] particular utterance is very large. The number which will be accepted as plausible by an informed audience, however, will be constrained by the particular context which is brought to bear upon the reading of the utterance' (Latour and Woolgar 1979, 35).

To argue for this they also invoke the concept of *circumstance*. This concept played a prominent role in a number of texts in Serres's *Hèrmes* series. Around the same time, Serres also gave courses on this topic in the United States, dealing with the role of 'circumstances' in grammar, logic, physics, thermodynamics and literature. As Serres would state in a later interview, 'the word "circumstance" was a bit my motto at that time' (Serres 2014, 91). In print it is first introduced as:

> What is around, what circles or surrounds something, without completely depending on it, this is the *circum*. Now all this is standing around, *stans*. The circumstance is at first spatial, and it is the drift in question. It is, on the other hand, mechanical: a certain equilibrium, fragile and temporary, which one understands, by hearsay, as a kind of constancy, but which isn't one. (Serres 1977a, 227)

The central idea is that a local order or stability is constituted and maintained by the things surrounding it and to which it stands in a relationship. 'Constancy is an exceptional formation on the non-ordered reality of circumstances' (Serres 1977a, 230). Due to these relationships there is a certain temporary stability. 'This is what the concept of circumstance expresses. It is neither stable nor instable. It is quasi-stable and quasi-instable in and by instability. It holds because it does not hold. And, because of this, it produces' (Serres 1977a, 228). What does it produce? Novelty, something that was not found in the circumstances. This perspective is picked up by Latour and Woolgar and applied to the construction of scientific facts, such as that of the Thyrotropin-releasing hormone TRF(H) in *Laboratory Life*:

> Circumstances (that which stands around) have generally been considered irrelevant to the practice of science. Our argument could be summarized as an attempt to demonstrate their relevance. Our claim is not just that TRF is surrounded, influenced by, in part depends on, or is also caused by circumstances; rather, we argue that science is entirely fabricated out of circumstance. (Latour and Woolgar 1979, 239)

It is in this sense that the argument becomes plausible that a statement cannot be 'verified outside the laboratory since the very existence of the statement depends on the context of the laboratory', since it is the combination of circumstances, 'the very network of social practice[,] which makes possible its existence' (Latour and Woolgar 1979, 183).

This reading sheds a similar light on their famous claim that the concept of reality 'cannot be used to explain why a statement becomes a fact, since it is only after it has become a fact that the effect of reality is obtained' (Latour and Woolgar 1979, 180). This claim would be misunderstood if it would be only read through the social constructionism idiom. Rather, the point is that reality cannot serve as an explanation since it would merely refer to the underlying disorder, whereas the scientific facts are not to be equated with these, but rather follow from a constructive process that stabilizes this initial disorder into order.

It is therefore warranted to look at the process of stabilization to account for the scientific facts, rather than to the original noise of which nothing meaningful could be said (unless one would already try to stabilize it). To illustrate this, they draw some parallels. One is in reference to Monod's *Le hasard et la necessité* (1970), which likewise argues that biological '[r]eality is constructed out of disorder, without the use of any preexisting representation of life' (Latour and Woolgar 1979, 250). Creating a scientific fact is similar to the process by which a new organism comes into being: that is a slow process by which initial disorder is (temporarily) stabilized into a fixed form. The second parallel is to Maxwell's demon:

> Maxwell's devil provides a useful metaphor for laboratory activity because it shows both that order is *created* and that this order in no way preexists the devil's manipulations. Scientific reality is a pocket of order, created out of disorder by seizing on any signal which fits what has already been enclosed and by enclosing it, albeit *at a cost*. (Latour and Woolgar 1979, 246)

So this second illustration also shows something else. It points to the necessary costs and interventions required to establish information, and thus scientific knowledge, and how this is a thoroughly materialist claim. As they explain in a footnote, Brillouin

> has made important contributions to a materialist analysis of science production. He regards *all* scientific activity (including the so-called 'intellectual' or 'cognitive' ones) as material operations which are in any way homologous to the usual object of physics. Since he provides a bridge between matter and information, he

also bridges the gap – so dramatic for the study of science – between intellectual and material factors. (Latour and Woolgar 1979, 260n18)

We thus end up in a materialist understanding of information. It is in this context that Latour and Woolgar link back to the notion of inscription: 'Our argument there was that writing was not so much a method of transferring information as a material operation of creating order' (Latour and Woolgar 1979, 245). In the next part, I wish to show how a similar understanding is also applicable to the concept, later picked up by Latour, of translation.

A sociology of translation

Translation is a central concept in what is typically known as ANT. Originally, it is derived from the work of Serres (1974a). Serres indeed starts that book by noting that we only know things through the system that transforms them, exemplified by the process of deduction in mathematics, induction in science, production in practical fields and translation (*tra-duction*) in the domain of texts (Serres 1974a, 9). Since Serres proposes to interpret all scientific activities as transformative acts on information, it is no surprise that translation becomes the general term for all these transformations. In the book Serres is mainly interested in the question of stability: To what extent are things kept stable throughout these procedures of transformation?

Latour does not use this concept in his early publications. It only becomes central later on due to the influence of Michel Callon, who around the time was picking up Serres's philosophy in a similar fashion. Callon was linked to the *Centre de sociologie de l'innovation* (CSI, created in 1967) at the *École des Mines*, an engineering school in Paris. Latour started to collaborate with Callon during his work on *Laboratory Life* and would become professor at the CSI in 1982. Later, Latour would recall how they together attended Serres's classes:

> We attended Serres's seminar every Saturday, in the smoke-filled amphitheater in the Sorbonne (people smoked in classrooms then!), profiting every time from the boldness with which Serres developed his 'anthropology of the sciences' based on the very fertile principle of exegesis according to which the single metalanguage of a text – a poem, a fable, a memoir, or a scientific treatise, it hardly mattered – could always be found in the text itself. All one had to do was look for it, a lovely methodological lesson for following the 'actors themselves' and an approach compatible with both semiotics and ethnomethodology. (Latour 2013b, 293–4)

Although Latour would only pick up the term in the 1980s, Callon already used Serres in an article in 1975 (Callon 1975). He would later define this approach as a 'sociology of translation' and apply it to a number of case studies, such as the electrical car (Callon 1981a, b) and fishermen and scallops (Callon 1984). Callon explicitly states that he 'owe[s] the concept of translation to M. Serres, *Hèrmes III, La traduction*' (1981a, 219n16). The concept of translation evokes a number of elements, such as that of displacement and that of becoming a spokesperson:

> To translate is to displace: the three untiring researchers attempt to displace their allies to make them pass by Brest and their laboratories. But to translate is also to express in one's own language what others say and want, why they act in the way they do and how they associate with each other: it is to establish oneself as a spokesman. (Callon 1984, 223)

Or, as Callon (1986) summarized elsewhere, translation consists in three components: that of a translator-spokesperson, where one actor starts to speak in the name of the many; the role the translator-strategist, where one actor convinces the others that one is an obligatory passing point; and the role of translation as displacement, where there is an active social and material intervention in the world to strengthen its position and to make the other elements behave in the desired way. Such displacement is always crucial according to Callon:

> Entities are converted into inscriptions: reports, memoranda, documents, survey results, scientific papers.... There are also movements of materials and of money. Translation cannot be effective, i.e. lead to stable constructions, if it is not anchored to such movements, to physical and social displacements. (Callon 1986, 27)

Latour takes up this concept in a 1981 paper written with Callon on the question of how to combine a micro- and a macro-analysis in sociology. Their solution is mainly to claim that macro-actors in the traditional sense (capitalism, culture, etc.) do not exist, but are the product of a translation process by which micro-actors become big: 'macro-actors are micro-actors seated on top of many (leaky) black boxes' (Callon and Latour 1981, 286). For this, they once again mobilize the concept of translation:

> By translation we understand all the negotiations, intrigues, calculations, acts of persuasion and violence, thanks to which an actor or force takes, or causes to be conferred on itself, authority to speak or act on behalf of another actor or force: 'Our interests are the same', 'do what I want', 'you cannot succeed without

going through me'. Whenever an actor speaks of 'us', s/he is translating other actors into a single will, of which s/he becomes spirit and spokesman. (Callon and Latour 1981, 279)

This project of Callon and Latour, later joined by authors such as Madeleine Akrich and John Law, was initially known in France as a sociology of translation (Akrich, Callon and Latour 2006). Hélène Mialet, a PhD student of Latour, reports how in the early 1990s, 'I never knew what Latour was referring to when he began talking about "actor-network theory", since I had known it as *La sociologie de la traduction*' (Mialet 2012, 457).

Indeed, around 1985 Callon started to use a second term to define their approach, namely that of 'actor-network' (Callon 1986, 1987). In his 1986 book, we find 'actor-network' in the glossary, with the following justification: 'The actor who speaks or acts with the support of these others also forms a part of the network (see translation centre). Hence the term actor-network, for the actor is both the network and a point therein' (Callon et al. 1986, xvi).

In his own chapter of the book, entitled *The Sociology of an Actor-Network*, Callon stresses three concepts to capture what he calls the 'co-evolution' of society, technology and science: 'the actor-world, translation and the actor-network' (Callon 1986, 20). Actor-world and actor-network

> draw attention to two different aspects of the same phenomenon. The term actor-world emphasizes the way in which these worlds, built around the entities that create them, are both unified and self-sufficient. The term actor-network emphasizes that they have a structure, and that this structure is susceptible to change. Accordingly, in later chapters the two are used interchangeably. (Callon 1986, 33)

Latour himself initially resisted the term, mainly because '[w]ith the new popularization of the word network, it now means transport *without* deformation, an instantaneous, unmediated access to every piece of information. That is exactly the opposite of what we meant' (Latour 1999d, 15). Only later Latour embraced the name (Latour 2005b).

Independently of the label ANT, Latour started using the concept of translation in his own texts from the mid-1980s onwards. It is present in his case studies on Pasteur, describing how '[t]he translation that allows Pasteur to transfer the anthrax disease to his laboratory in Paris is not a literal, word-for-word translation. He takes only one element with him, the micro-organism, and not the whole farm, the smell, the cows, the willows along the pond or the farmer's pretty daughter' (Latour 1983, 146).

The influence of Serres is especially present in his book on Pasteur, which is even dedicated to him, acknowledging at multiple places that the 'book owes a great deal to Michel Serres' work' (Latour 1984, 251n2). In a similar way to Callon, Latour stresses the different aspects of translation:

> First translation means drift, betrayal, ambiguity . . . . It thus means that we are starting from *inequivalence* between interest or language games and that the aim of the translation is to render two propositions equivalent. Second, translation has a strategic meaning. It defines a stronghold established in such a way that, whatever people do and wherever they go, they have to pass through the contender's position and to help him further his own interests. Third, it has a linguistic sense, so that one version of the language game translates all the others, replacing them all with 'whatever you wish, this is what you really mean'. (Latour 1984, 253n16)

The book also draws from another book by Serres, *Le Parasite* (1980), in which Serres uses the figure of the 'parasite' to sketch his take on the logic of relations (see Chapter 8). According to Serres, any kind of relation, by its nature, is defined by distortions, and thus as a parasite. 'There is no system without parasites' (Serres 1980, 12). Whenever something is translated from one medium into another (e.g. chemicals in a graph; French in English; social phenomena in statistics) something is distorted, added, left out and transformed. If there were no such distortions, the translation would be invisible, since no difference between the two would be manifest. Most human interventions fight off these distortions by making them less extreme, less visible or more acceptable. But getting rid of all distortions is impossible. Or, in the words of Latour: 'Nothing is, by itself, the same as or different from anything else. That is, there are no equivalents, only translations. . . . If there are identities between actants, this is because they have been constructed at great expense' (Latour 1984, 162).

Latour applies this logic to the case of Louis Pasteur, mainly in order to understand how Pasteur was capable to recruit allies by translating their interests, problems and identities in his network. Latour claims that the Pasteurians have not just freed us from the 'parasites' that distorted our ordinary social life, the microbes. What they rather did was merely to translate one parasite into another, namely themselves:

> Microbes are everywhere third parties in all relations, say the Pasteurians. But how do we know this? Through the Pasteurians themselves, through the lectures, the demonstrations, the handbooks, the advice, the articles that they produced from this time. . . . Serres describes this elimination of a parasite by

another more powerful one. Only after the insulation of the second parasite can we declare ourselves safe from the first. At the cost of setting up new professions, institutions, laboratories, and skills at all points, we will obtain properly separated channels of microbes, on the one hand, and of pilgrims, beer, milk, wine, schoolchildren, and soldiers, on the other. (Latour 1984, 38–9)

While the doctors first become the spokespersons of microbes in their laboratories in order to translate the concerns of these actors about health, hygiene or warfare into their own ('If you want a healthy population that can serve in your army, you need to focus on microbes'), they also introduce new distortions, since laboratories have to be funded, hands have to be washed or cows have to be vaccinated. As a result, French society has radically altered, since from now on it includes well-defined 'microbes and microbe-watchers in its very fabric' (Latour 1983, 158).

Translation and inscription

We have seen how Latour mobilized both the concept of inscription, as developed by Dagognet, and that of translation, as developed by Serres. I want to end by arguing that Latour's originality lies in the fact that he started to combine both these concepts, and made them work together, focusing on how inscriptions are transported and thus translated to different settings. To focus on inscriptions on paper only, for instance, would lead one to argue for 'a mystical view of the powers provided by semiotic material – as did Derrida', whereas to focus solely on translations of interests and groups 'would be to offer an idealist explanation (even if clad in materialist clothes)' (Latour 1986, 6). Rather, Latour's aim 'is to pursue the two lines of argument at once. To say it in yet other words, we do not find all explanations in terms of inscription equally convincing, but only those that help us to understand how the mobilization and mustering of new resources is achieved' (Latour 1986, 6).

The answer is found by focusing on the question of *mobilization*. In this context Latour coins the term 'immutable mobiles', that is 'you have to invent objects which have the properties of being *mobile* but also *immutable, presentable, readable* and *combinable* with one another' (Latour 1986, 7). Without such immutable mobiles, modern science becomes impossible. In the history of science, elements such as 'organized scepticism, scientific method, refutation, data collection, theory making' are insufficient, since they all previously 'had been tried, and in all disciplines: geography, cosmology, medicine, dynamics, politics, economics and so on. But each achievement stayed local and temporary just because there was no way to move their results elsewhere and to bring in those of others without new corruptions or errors being introduced' (Latour 1986, 12). Only by combining inscription and translation in the concept of mobilization can a meaningful explanation be proposed:

> it is not the inscription by itself that should carry the burden of explaining the power of science; it is the inscription *as the fine edge* and *the final stage* of a whole process of mobilization, that modifies the scale of the rhetoric. Without the displacement, the inscription is worthless; without the inscription the displacement is wasted. This is why mobilization is not restricted to paper but paper always appears at the end when the scale of this mobilization is to be increased. Collections of rocks, stuffed animals, samples, fossils, artifacts, gene banks, are the first to be moved around. (Latour 1986, 17)

The act of translation, especially in the sense of mobilization, is thus always a material activity. We have simply forgotten the material side, since it has become so obvious. 'We are so used to this world of print and images, that we can hardly think of what it is to know something without indexes, bibliographies, dictionaries, papers with references, tables, columns, photographs, peaks, spots, bands' (Latour 1986, 14).

Conclusion

This chapter started from a paradoxical reception of Latour's work, being both understood as part of a tendency to reduce everything to texts and simultaneously as part of a new realist movement. Through an analysis of Latour's connections with the work of Dagognet and Serres I argued that this paradox is not a sign of confusion or a weakness of Latour's work, but rather an explicit starting point: focusing on texts and focusing on objects is not so different since texts have a materiality too. In that sense, texts – or better inscriptions – can be understood in a material way. This equation of word and world, however, does not occur in the traditional sense, namely that of a reduction of things to their descriptions, but rather in the opposite direction. All texts are to be understood in a material way, just as other objects.

In 1970s French philosophy we thus find a distinctive object-oriented programme, that subsequently influenced Latour and ANT. Later work of Latour consisted thus not in a 'turn to realism' as some have argued, that allegedly breaks with his earlier text-centred work (e.g. Bourdieu 2004, 29). It is rather a continuation –

or, more adequately, a translation – of a number of insights of Dagognet and Serres. In the work of both these authors, we see such an explicit object-oriented focus, which – similar to Latour – only became more explicit in later work.

In the case of Dagognet this is clear from a number of publications in the 1980s, especially *Rématerialiser* (1985) and *Éloge de l'objet* (1989). In these books he argues that philosophy has neglected materials and objects, in favour of symbols and the subject. This has also been noted by a number of commentators (e.g. Beaune 2011). Christian Godin, for instance, argues that 'Dagognet openly declares himself hostile to hermeneutics, which replaces the real by texts and objects by the subject' (Godin 2011, 83). Dagognet aims to rebalance this, both by an analysis of a number of new materials made in science and technology, and by 'rematerialisering the history of philosophy' (Dagognet 1985, 182). A similar message is found in *Éloge de l'objet*, which starts from the idea 'that the object, for which we sing the eulogy here, has not attracted, in the so-called consumer society, the attention, much less the favor of the philosophers. They are wary of it. They have been more attracted to the subject' (Dagognet 1989, 9). Dagognet analyses a number of concrete materials, such as glass, in order to

> draw the object from the shadow, show its dignity, importance, complexity. How can one not want then, by all means, that all can benefit from it? Thus, as will not surprise anyone, this involves an additional redefinition of social connections. (Dagognet 1989, 228)

This object-oriented approach is present in Serres's work as well, from his *Hèrmes* series onwards. By recognizing that all objects emit information, and thus a symmetry between humans and non-humans can be maintained, objects play a crucial role again. Serres's ambition is '[t]o return to the things themselves, to the mixed multiplicities, to the dispersions by taking them as they are' (Serres 1977a, 40) and 'to free oneself from the prison of words, the walls of screens, effects of illusion. Go out. The renaissance, once again, after scholasticism, of a philosophy of nature. The things themselves. Yes, the return to materialism, a forgotten materialism' (Serres 1977a, 156–7) that the ambition even becomes stronger in the 1980s, especially in *Les cinq sens* (1985), *Statues* (1987) and *Le contrat naturel* (1990). In each of these books Serres pleads to go beyond language and the social and to acknowledge the role that objects play. Already in *La distribution* he suggests the meaningfulness of a 'transcendental object' (Serres 1977a, 58), that is that society and the subject are constituted by things. These transcendental objects are what Serres will start calling *quasi-objects* (see Lehtonen 2020):

> I imagine, at the origin, a rapid vortex in which the transcendental constitution of the object by the subject would be fueled, as though feeding back, by the symmetrical constitution of the subject by the object, in lightning-fast semi-cycles and ceaselessly repeated, coming back to the origin. . . . There is an objective transcendental, the constitutive condition for the subject through the apparition of the object as object in general. We have testimony of the converse or symmetrical condition on the eddying cycle, traces or narratives, written in the labile languages . . . . But we have tangible, visible, concrete, fearsome, silent witnesses of the direct constitutive condition starting from the object. However far back we may go into the talkative history or the silent pre-history, they are always there. (Serres 1987, 119)

A similar message is found in Latour's work, most famously in his call for a parliament of things (see Chapter 8). But also in lesser-known articles, Latour criticizes a tendency in philosophy to reduce materialism to a very narrow-minded and even idealist conception of matter: 'How could they have confused *res extensa* and materiality when every object, every artisan, every skilled gesture would have told them the very opposite' (Latour 2007, 132)? Moreover, he starts using the concept of quasi-object, drawn from Serres's book *Statues* (1987), a book Latour reviews in 1990, together with Steven Shapin and Simon Schaffer's *Leviathan and the Air-Pump* (1985), laying the foundation of Latour's book *Nous n'avons jamais été modernes* (1991).

Especially *Leviathan and the Air-Pump* proved crucial for Latour, since the authors come so close to what Serres is also trying, namely to 'make their analysis and that of their characters turn *around the object,* around *this* specific leaking and transparent air pump' (Latour 1990a, 152). In that sense, the concept of inscription and translation, combined as mobilization, allows for a view of science and society in which there is room for the transcendental object:

> There is nowhere to be seen an object and a subject, a primitive and a modern society; There are only series of substitutions, of displacements, mobilizing people and things on larger and larger scale and size. Serres imagines a spiral, each loop of which represents a co-production of a collective, and of an object by the displacement of one social entity by another one which is more non-social, more thing-like. (Latour 1990a, 163)

In that sense, behind much of Latour's work – and even ANT in general – there is a concrete and consistent metaphysical programme in which the world is interpreted in a relational way, where all objects are connected and in interaction with one another. Objects themselves, moreover, are seen as only temporary stabilized entities that only persist as long as their stability is actively maintained. From this perspective, central insights from ANT, such as the generalized symmetry, where agency is distributed among humans and non-humans, and its constructivist interpretation of facts and artefacts, are not surprising consequences. To fully understand these claims, it is necessary to pay attention to a number of French philosophers, from which Latour and others draw inspiration, such as Michel Serres, but also François Dagognet. Only then it becomes possible to see how, what first might look like a paradox, is actually a coherent project. In the next chapters we will further explore this through the themes of time and modernity (Chapter 5), anthropology (Chapter 6), religion (Chapter 7) and ecology (Chapter 8).

5

Brewers of time: Michel Serres and modernity

Introduction

We have seen in the previous chapters how Serres developed an alternative normative take on what it means to do science (Chapter 3), inspired by a different ontological view of the world and of science (Chapter 4). This brings us to the following question: If the world is indeed to be analysed as a network, how must we situate ourselves in it, in both space and time? The previous chapters thus bring us to another important dimension of Serres's philosophy: his analysis of our modern condition and questions it raises concerning history and temporality.

That the notion of a quasi-object offers us an alternative way to look at our modern selves has not escaped Bruno Latour. As we saw at the end of the previous chapter, Latour, faced with a world full of entities that can be classified as neither natural nor social, invokes the help of Michel Serres: 'we may now be able to locate the position of these strange new hybrids and to understand how come that we had to wait for science studies in order to define what, following Michel Serres (1987), I shall call quasi-objects, quasi-subjects' (Latour 1991, 51). Remarkably, quasi-objects also challenge the way we look at time: 'The proliferation of quasi-objects has exploded modern temporality along with its Constitution' (Latour 1991, 73).

To fully understand this matter, it is crucial to understand what Serres and Latour are opposing. The most obvious candidate is 'modernity', against which Latour seems to position his own and Serres's philosophy. Nonetheless, I believe it is essential to consider another popular candidate by which philosophers have positioned themselves: *postmodernity*. There are several reasons for this. First of all, their work is often classified and discussed as part of postmodernism. Secondly, the diagnosis of postmodernity was very influential at the time when

Serres and Latour were formulating their alternatives and they actively distanced themselves from it (e.g. Latour 1990).

Most importantly, there is a direct connection with Jean-François Lyotard, the philosopher who popularized the term 'postmodernity', defining our postmodern condition 'as incredulity towards metanarratives' (Lyotard 1979, xxiv). Though it has hardly been noticed (however see Schmidgen 2014, 4), there are direct indications for such a connection. First of all, both Serres and Latour were in the United States in the mid-1970s, at that time a hotspot for 'French Theory'. Latour was doing his fieldwork for *Laboratory Life* at the Salk Institute and regularly visited the campus in San Diego, where Lyotard was teaching. Latour picked up Lyotard's notion of 'the agonistic' from the latter's lectures on Nietzsche, decadence and the sophists, which Latour attended (Schmidgen 2014, 34).[1] A key intermediate figure was Paolo Fabbri, an Italian semiotician and a regular visitor at the University of California in the 1970s. While Latour published his first papers with Fabbri (Latour and Fabbri 1977), the latter was also a good friend of Lyotard, often mentioned in his work (e.g. Lyotard 1983, 14, 17, 129; 1986, 13–14).

Conversely, Lyotard's *La Condition postmoderne* (1979) supports its claims about science with plenty of references to French philosophy of science. He refers not only to Bachelard or Canguilhem but also Serres's *Hèrmes* series is regularly referred to, in support of Lyotard's claim that we live 'in a society whose communication component is becoming more prominent day by day, both as a reality and as an issue' (Lyotard 1979, 16). Lyotard also refers to Serres's idea (1974a, 94) that '[i]nformation itself costs energy, and the negentropy it constitutes gives rise to entropy' (Lyotard 1979, 99). The book is also remarkable for containing one of the first references to the work of Latour, namely the early paper with Fabbri (Lyotard 1979, 24).

Implicit references to Latour and Serres are also present in later works by Lyotard. We will come back to his references to Serres, which will be crucial. But also Latour seems often implicit at play (e.g. Lyotard 1986, 62, 75). An explicit mention of Latour is found in *Le Différend*, in the context of a *différend* concerning the debates surrounding the possibility and the meaningfulness to analyse science sociologically, for instance by a focus on the rhetoric of science (Lyotard 1983, 17). Lyotard refers here to Latour's then-still unpublished manuscript of what will later become *Irreductions,* the second part of *The Pasteurization of France* (1984). Both scholars seem to share the ambition to overcome the opposition between right and might, rationality and power, through the Nietzschean concept of force.

On his turn, Latour refers to Lyotard, but in an ambiguous way. On the one hand, he is quite positive about Lyotard's work:

> Actually I like *The Differend*. I will pass politely over his radical past; he wrote a whole series of absolutely awful books, but then he did some quite interesting work when he examined the origin and distinction between right and might in his argument on sophistics. That work rehabilitated sophistics against the *coup de force* of Plato. I find that book influential, and his Great Narrative stuff is also useful. (Latour and Crowford. 1993, 254)

On the other hand, Latour is very critical of Lyotard and postmodernism in general, stating, 'I have not found words ugly enough to designate this intellectual movement' and '[t]here is only one positive thing to be said about the postmoderns: after them, there is nothing' (Latour 1991, 61–2).

Such criticisms, however, are not surprising given that Latour and Serres's projects pose great challenges to Lyotard's diagnosis: How can we be postmodern if we have never been modern in the first place? And, similarly, the 'so-unmodern work' (Latour 1991, 84) of Serres seems to mock Lyotard by introducing the notion of the 'Great Story' [*Grand Récit*] of humanity, going from the Big Bang to our contemporary world: 'This Great Story itself does only date from a few years, at the same moment when, by an irony of ignorance, philosophers characterized the present times, said by them to be postmodern, by the disappearance of great stories' (Serres 2016, 14–15).

I believe it therefore to be more revealing to contrast Serres's understanding of temporality and history, further developed by Latour, with that of postmodernity. As I hope to show, both views share a great number of common themes. Therefore, I will start with Lyotard's position and Serres's and Latour's criticisms. Then I will elaborate on Lyotard's own position in order to subsequently highlight what I believe is really at stake, namely the challenge of performativity which Lyotard sees as problematic. Through a confrontation with Lyotard's work we will be able to see better what is ethically and politically at stake in Serres's philosophy of quasi-objects: the question of the qualification of relations. If the world consists of relations, this at first seems to dismantle all possibilities of critique: nothing is given, everything is constructed. What would then be the ground for any meaningful distinction between a desired and undesired construction, for relations to be cultivated and relations to be avoided? This conversation with Lyotard will put us on track to find an answer to these questions, which we will take up in the next three chapters.

Modernity and postmodernity

Jean-François Lyotard is mostly known for *La condition postmoderne* (1979), a report on the status of knowledge in Western industrialized societies, commissioned by the Conseil des universités du Québec. To Lyotard's surprise it became an international bestseller. Nonetheless, Lyotard's oeuvre is more diverse. My ambition is not to give a full overview of his work (see Williams 1998), but just to sketch the main points: after a political phase dominated by Marxism, Lyotard started to write more philosophical texts. From the 1960s on Lyotard's work became very critical of structuralism's exclusive focus on language and structures, as it ignored what is called by him 'the figural', exemplified in art: sensuous forms which cannot be reduced to structure (Lyotard 1971). *L'économie libidinale* (1974), a book to which we will return, aimed to analyse society as a network of libidinal energies, that regularly petrify in structures (such as bodies or institutions), but always tend to break free or partly escape from them.

His work on postmodernity is, however, part of another shift in his philosophy in the 1970s under the influence of the Immanuel Kant and the 'linguistic turn', mainly drawing on Ludwig Wittgenstein. He starts to stress the value of heterogeneity and incommensurability between different genres of discourse. Initially, Lyotard framed his position not as postmodernism, but as *paganism* (Lyotard 1977a, 1977b; Lyotard and Thébaud 1979). It is this work on paganism and the sophists that Latour also praises and refers to. To schematize, paganism can be characterized by (at least) seven elements: (1) paganism starts from the fact that there is no ultimate God, no metanarrative through which one can judge all forms of speaking and acting. It is 'the denomination of a situation in which one judges without criteria' (Lyotard and Thébaud 1979, 16). Lyotard often opposes this to Plato, whose philosophy centres around the belief in an objective, transcendent criterion by which to order particular situations. (2) The fact that no such metanarrative is available does not imply the impossibility of judgement, but only that the nature of judgement is different. A judgement does not need an ultimate ground. Instead, Lyotard uses several alternative models, ranging from Nietzsche's will to power, Aristotle's notion of *phronesis* to Kant's notion of the imagination, seen as 'a power to invent criteria' (Lyotard and Thébaud 1979, 17). There is thus a demand for criteria, albeit one that can never be ultimately met, but must be answered and created locally and provisionally (depending on the god one is dealing with). (3) Paganism starts from an irreducible heterogeneity: a polytheism. Paganism is not atheism. There are still hierarchies. (4) Given polytheism, the strategy to engage with the world is not one of deduction (from the principle), but of seduction, where humans try to alter the behaviour of the gods, to 'come to terms with them by means of counter-tricks, offerings, promises, small contracts of alliance giving rise to complicit ceremonies, all of this in humor and fear' (Lyotard 1977a, 43). (5) In this practice of seduction, no speaker is ever fully autonomous. This does not imply a fatalism. 'It is precisely because they are not the authors of meaning that they must always rely on their wits. We are never in the autonomy of power' (Lyotard and Thébaud 1979, 36). (6) In addition, there is no perfect control over the effects of one's actions, no guarantee that the intended effect will be achieved. It is this uncertainty and ambiguity that philosophers such as Plato fear, leading them to replace this model of seduction with one that offers full control (Lyotard and Thébaud 1979, 4). (7) Therefore a pagan attitude is one of *humour* rather than *irony*. Irony entails that one denounces a certain point of view from another, higher point of view: I ironically see the futility of their point of view, which is nothing but an illusion. Humour, on the other hand, is capable to 'point out the ruse at the very site of its outpost and without having to go to another outpost to do it. This is what the gods of antiquity did when one of them claimed to be the only true god' (Lyotard 1993, 83).

This notion of paganism will transform into the notion of the postmodern (Lyotard and Thébaud 1979, 16) and *La condition postmoderne* (1979) links this with the question of the contemporary legitimation of knowledge: 'who decides what knowledge is, and who knows what needs to be decided?' (Lyotard 1979, 9). Scientific or cognitive statements cannot be legitimated on their own, since they only tell what is the case, not what should be the case. Such a legitimation, however, is typically provided in modernity by a metadiscourse:

> I will use the term *modern* to designate any science that legitimates itself with reference to a metadiscourse of this kind making an explicit appeal to some grand narrative, such as the dialectics of Spirit, the hermeneutics of meaning, the emancipation of the rational or working subject, or the creation of wealth. (Lyotard 1979, xxiii)

In contrast, postmodernity is interpreted 'as incredulity towards metanarratives' (Lyotard 1979, xxiv). Lyotard mainly discusses two such metanarratives: the metanarrative of the emancipation of the rational subject (linked with the Enlightenment) and the metanarrative of the history of the universal spirit (linked with the speculative philosophy of German Idealism) (Lyotard 1979, 31–7). Lyotard situates the shift to postmodernity around the Second World War, when these metanarratives start losing their legitimizing power and are replaced by a new criterion for knowledge, the criterion of performativity. At the same time, however, Lyotard pins his hope on an alternative way of doing science, found within science itself, that he links with the notion of paralogy: the practice of looking for instabilities, that change the whole system and its rules, invent new rules, similar to the postmodern artist.

Lyotard's hypothesis is that these metanarratives lost their plausibility because we are witnessing 'an internal erosion of the legitimacy principle of knowledge. There is erosion at work inside the speculative game, and by loosening the weave of the encyclopedic net in which each science was to find its place, it eventually sets them free' (Lyotard 1979, 39). Here we enter Lyotard's most important book, *Le Différend* (1983): these metanarratives lost their plausibility due to the overabundance of new differends that fail to go away.

Lyotard defines a differend in opposition to a litigation: 'a differend [*différend*] would be a case of conflict, between (at least) two parties, that cannot be equitably resolved from lack of a rule of judgment applicable to both sides of the argument. One side's legitimacy does not imply the other's lack of legitimacy' (Lyotard 1983, xi). The most famous example Lyotard develops is that of Robert Faurisson and Holocaust denial:

> His argument is: in order for a place to be identified as a gas chamber, the only eyewitness I will accept would be a victim of this gas chamber; now, according to my opponent, there is no victim that is not dead; otherwise, this gas chamber would not be what he or she claims it to be. There is, therefore, no gas chamber. (Lyotard 1983, 3–4)

But Lyotard gives other examples as well: a publisher who asks you to name a book that is refused by all publishers and yet is important; a communist regime who argues that any real communist would acknowledge that the state has the authority to decide what communism should look like, so only a non-communist would criticize the state (Lyotard 1983, 4); the worker who cannot contest that his labour is a commodity, for any code of law starts from this assumption (Lyotard 1983, 10); the animal who cannot contest the violence done to it, since no court acknowledges it as a subject (Lyotard 1983, 28).

That such differends come into being follows from the earlier described paganism: there is no universal metanarrative through which all events can be grasped. 'Phrases obeying different regimes are untranslatable into one another' (Lyotard 1983, 48). Rather events always (partly) escape from our attempts to grasp them through our ways of speaking. What has happened in the twentieth century is that a number of such events occurred that produced numerous differends for all the available metanarratives:

> Everything real is rational, everything rational is real: 'Auschwitz' refutes speculative doctrine. This crime at least, which is real . . ., is not rational. – Everything proletarian is communist, everything communist is proletarian: 'Berlin 1953, Budapest 1956, Czechoslovakia 1968, Poland 1980' (I could mention others) refute the doctrine of historical materialism: the workers rose against the Party. – Everything democratic is by and for the people, and vice-versa: 'May 1968' refutes the doctrine of parliamentary liberalism. The social in its everydayness puts representative institutions in check. – Everything that is the free play of supply and demand is favorable for the general enrichment, and vice-versa: the 'crisis of 1911 and 1929' refute the doctrine of economic liberalism. And the 'crisis of 1974–1979' refutes the post-Keynesian revision of that doctrine. The passages promised by the great doctrinal synthesis end in bloody impasses. Whence the sorrow of the spectators in this end of the twentieth century. (Lyotard 1983, 179–80)

What is important in this claim is that Auschwitz or May 1968 are first of all events linked to a *proper name*, rather than a fixed description. Lyotard is here influenced by Saul Kripke's *Naming and Necessity* (1980): in contrast to what authors such as Bertrand Russell thought, proper names gain their meaning not so much from being abbreviations of descriptions (e.g. 'Saul Kripke is the author of *Naming and Necessity*'), but are rather the product of an initial baptismal event, forming the origin of a causal chain of referential language use up to its current users. Lyotard uses this insight to argue that one can refer to events without relying on any fixed descriptions of them. We can talk about Auschwitz, even refer to it properly, without having to define what it is and what it stands for.

Such proper names, Lyotard argues, are the basis of our culture: one enters a culture by learning its proper names, its heroes, places and important historical dates: all rigid designators in the sense of Kripke. The names we are faced with in our postmodern condition, however, 'have the following remarkable property: they place modern historical or political commentary in abeyance' (Lyotard 1989, 393). The role these proper names play in the liquidation of our metanarratives is that they systematically fail to be recuperated in any of the available metanarratives. Thus Auschwitz, for instance, imposes itself as a proper name and though we are unable to articulate its full meaning, it nonetheless distorts all our attempts to fit it in to a greater narrative.

How to resolve this? According to Lyotard the options are limited. The only way to solve a differend is by providing it with a (new) genre of discourse: 'What is at stake in a literature, in a philosophy, in a politics perhaps, is to bear witness to differends by finding idioms for them' (Lyotard 1983, 13). There is, however, no guarantee that such an idiom can be found or would not result in new differends. Again a pagan attitude, which aims for local and temporary solutions, is the best we can do.

Serres's Great Story of the Universe

At the beginning of the twenty-first-century Serres wrote four books, which he saw as constituting a series: *Hominescence* (2001), *L'Incandescent* (2003a), *Rameaux* (2004) and *Récits d'humanisme* (2006). The thread through these books is what Serres calls the Great Story of the Universe. Serres finds it ironic that the sciences have provided us with this story, at the end of the twentieth century, precisely at the moment when 'philosophers characterized the present times, said by them to be postmodern, by the disappearance of great stories' (Serres 2016, 15). He clearly has Lyotard in mind.

This last quotation dates from one of Serres's final books, on the philosophy of history. Serres starts that book by a critique of traditional historiography: 'Historians readily boast of exercising, practicing, celebrating memory, while their discipline is defined more as a series of forgetfulness' (Serres 2016, 7). This forgetfulness first of all refers to the fact that history typically starts with the invention of writing, thereby ignoring – in the past and present – those groups which do not write. But this is only the tip of the iceberg, since historiography also tends to focus on the human, thus forgetting the history of life and the universe.

The Big Story, on the other hand, is precisely characterized by the inclusion of these non-human dimensions. For Serres 'all things around us and in us – planets and beasts, bacteria, rocks and metals, Earth and Sky, the World and the Universe – have a story, as relatable as ours, conditional to ours, without which ours would not exist, but which the latter, obsessed with navel-gazing, forgot' (Serres 2016, 15). A focus on history and time becomes central to all sciences and as a result 'scientists become the historians of the Great Story' (Serres 2016, 27). The content of the Great Story is one most of us are familiar with, since it has become part of our education, but was revolutionary in the twentieth century: the story of the Big Bang and the formation of atoms; the story of the formation of the stars and our solar systems; the story of the origin of life and its subsequent evolutionary history; the story of *homo sapiens* and its emigration out of Africa; the story of the invention of agriculture and animal husbandry; and finally the story of the invention of writing and traditional history.

What, however, has made this Great Story possible? Certainly it is partly explained by specific developments in the individual sciences. But, more importantly, there is an underlying element that made this synthesis possible: the idea that the world is an informational network, written in languages we can decipher (see Chapter 4):

> Inert or alive, the universe speaks like us, writes like us, says and expresses itself like us, creates databases, remembers, translates, and even sometimes mutates, wanders or lies, but rarely. This Ptolemaic revolution of language decenters it from us. Will objective languages outnumber the idioms of men? Does nature relinquish our linguistic exclusivity? (Serres 2006, 80)

The result is a final blow to our narcissism, after Copernicus told us we are not the centre of the universe, Darwin revealed that we are not the centre of life and Freud showed us that consciousness is not master in its own house:

> What we said about ourselves – language, traces, symbols, signs and meanings, writing and memory, cognition and *cogito* – is taken on and demonstrated by all things in some way. The Grand narrative of vast time shunts the intermediary of memory through language. Hence the status of language and of our intelligence: intermediaries between the world and the world, between things and things, a sort of brief noise interference. (Serres 2003a, 35)

Nonetheless, Serres links this simultaneously with a number of shifts with regards to the position of humanity in the world, a process that he calls *hominescence*. Similar to adolescence, it is a process of transformation that humanity is going through, which, according to Serres, reaches a number of thresholds in the twentieth century, linked mainly with contemporary developments in biotechnology. The first event is the end of agriculture, in the sense that, whereas throughout history most people have always worked on the land, this has since the 1950s become a minority. For Serres, this 'is the greatest event of the 20th century' (Serres 2001, 72). The second event relates to our human bodies: similar to how we became more independent from the land, biotechnology made us independent from our bodily constitution. Through medicine and technology we learned to control our bodily processes: 'Exo-Darwinism is what I call this original movement of organs towards objects that externalize the means of adaptation. Thus, exiting evolution with the first tools, we entered into a new time, an exo-Darwinian one' (Serres 2001, 39).

As a result, we realize that our bodies and our humanity never resided in a fixed essence, but rather remain undefined and open. 'What is the body? Answer: It is not; it was, but is not any longer; for it now lives in the mode of the possible; only modal logic allows us to apprehend it; it leaves necessity to enter into the possible. This is the best definition that can be given of it: an incarnated virtual' (Serres 2001, 29). What constitutes a body, and even a human, is pure potentiality – the human is in essence totipotent or omnipotent:

> By losing countless specificities, valences or real powers, the zero-valent, nil-potent human became, without meaning to no doubt, virtually omnivalent, totipotent, global and infinite. These impoverishments disadapted it to every thin and precise local niche and left it with no bounds or definition. Undefined in a few organs as well as in our possibilities, we became the champions of inadaptation; we don't even know how to define ourselves. (Serres 2003a, 41)

This leads to two important consequences, clearly in opposition to Lyotard: a renewed humanism and universalism (see Watkin 2016). According to Lyotard any kind of grand narrative of humanism had become impossible. Serres concludes the opposite: only now do we have our first, genuine form of humanism. The old humanism was nothing but a particular history imposed by the West. But since the discovery of the Great Story we can tell our shared history of our origins in Africa and our exodus from it: 'Until today, indeed, humanism never took place because the universal human it referred to did not exist' (Serres 2006, 36). This new humanism, moreover, is similarly defined as the body: 'What is this humanity? Answer: a possibility in the range of potentialities, potency, yes, omnipotency since it can become anything. What is man? This range itself, this omnipotency' (Serres 2001, 48).

Secondly, Lyotard also denied the possibility of any form of universalism, since it would deny the heterogeneity of genres of discourse, and thus create differends. Again, Serres challenges this. Serres agrees that the traditional Western idea of universalism is problematic, since 'by excluding non-Europeans, other humans, all the living species, the inert planet and the world as a whole, these corporatist intellectuals, believing themselves to be alone in the world, practiced a universal racism' (Serres 2003a, 104). But the Great Story offers an alternative universalism:

> For the Grand Narrative truly invents globalization by virtue of the fact that it attaches itself, for the first time, to the fate of the entire Universe, of the entire

Earth, of all living things, of all humanity while only attaining politics and cultures at the end of the account and contingently, both of them minuscule in such a picture. (Serres 2003a, 104)

Universalism is thus dissociated from imperialism and the assumption that universality implies standardization and unity. Serres offers a number of existing non-violent and non-uniform universalities as 'counterexamples' (Serres 2003a, 209). First example: the weather. Despite following universal principles and laws, the weather is not uniform, but rather heterogeneous: 'These local differences without any uniformity are produced by the global climate system. So there exists at least one universal model with highly singularized local variants' (Serres 2003a, 210). Second example: the universe, similarly, creates endless different variations of solar systems and planets, despite following universal principles. Third example: life, which 'exploded on the basis of a single code formed from only four elements and their combinations; with time, countless species, individuals and varieties appeared according to selections, mutations, climates and latitudes' (Serres 2003a, 210). Fourth example: agriculture. 'Today, more than nine-tenths of our food comes from agrarian practices. And yet no culture has the same table manners nor the same cuisine, nor above all the same tastes' (Serres 2003a, 212).

These examples show a universalism that does not fall prey to all the troubles that Lyotard associates with it. For Serres, this universalism can in fact provide ammunition to fight the universal imposition of Western particularism, which imposes one particular way of organizing the world in a uniform manner: 'It's not a matter of setting the local against the global but, quite the contrary, of fighting with the global against this local' (Serres 2003a, 217).

In Serres we thus find what at first sight seems to be a very modern emancipatory project: humanity is freeing itself from the constraints of body and land to achieve a universal humanism. But such an assessment is too hasty. Serres's evaluation of these developments is not unambiguously positive. Agriculture, for example, was more than a constraint, but constituted a societal network that reciprocally enabled humans and non-humans to redefine themselves: 'Up until the aforementioned dates, practices and sciences, arts and religions, languages and cultures flourished with and due to agriculture' (Serres 2001, 71). Agriculture was a productive form of purification that enabled new collectives to be created on the cultivated fields of the farmers, subsequently shaping what our societies looked like.

At the same time, these new developments do more than just make us less dependent. A central theme of Serres is that this emancipation, paradoxically, also creates new forms of dependency (see Chapter 7). Serres deviates from the classical modern story by stressing how our liberation, our hominescence and exo-darwinism, is not just a product of subjects, but always also of quasi-objects around which our collectives are organized. Thus our independence of the land and of our bodies is only accomplished by mobilizing numerous other elements, such as the technologies required to maintain this independence. These technologies, on their turn, impact our world to the extent that it is threatened by them. 'We depend, in fact, from now on, on things which, quite rightly, depend on us, on the effects of our victories' (Serres 2006, 138).

Latour's we have never been modern

Another way to understand why Serres's Great Story does not follow a modern structure is by a detour through the work of Latour. We already saw how Latour was influenced by Serres's philosophy of quasi-objects. But Latour also mobilizes Serres's philosophy to challenge the idea of modernity itself. He follows Serres's criticisms of Bachelard and argues that to be modern is never merely just the practice of purifying our minds and our worlds, untangling the subjective from the objective. It always also includes another practice or dimension, namely the act of creating new hybrids or quasi-objects. This proliferation of hybrids, however, challenges our modern way of organizing our world, the modern constitution:

> the proliferation of hybrids has saturated the constitutional framework of the moderns. The moderns have always been using both dimensions in practice, they have always been explicit about each of them, but they have never been explicit about the relation between the two sets of practices. (Latour 1991, 51)

Latour challenges many facets of the modern constitution, but one that is particularly relevant here is its dual way of structuring time: 'The proliferation of quasi-objects has exploded modern temporality along with its Constitution' (Latour 1991, 73). What is specific to modern temporality, according to Latour, is the idea that the past can be abolished and forever surpassed:

> [The Moderns] do not feel that they are removed from the Middle Ages by a certain number of centuries, but that they are separated by Copernican revolutions, epistemological breaks, epistemic ruptures so radical that nothing of that past survives in them – nothing of that past ought to survive in them. (Latour 1991, 68)

The result is the particular situation in which certain objects, ideas or cultures, though spatially nearby, can be temporarily completely separated from us. These elements are seen as 'surpassed' or 'obsolete', made inoperative, yet stored in museums or preserved as a relic of a time we can never get back to.

Latour contests that this is how the past functions in our contemporary society, or more precisely: it does function in this modern way, but only because we have constructed this form of temporality. Similar to the purification of objects and subjects, the modern temporality is real, because it is constructed; but such a temporal purification presupposes a process of proliferation of quasi-objects, that consist not only of human and non-humans, but also of old and new things:

> The impression of passing irreversibly is generated only when we bind together the cohort of elements that make up our day-to-day universe. It is their systematic cohesion, and the replacement of these elements by others rendered just as coherent in the subsequent period, which gives us the impression of time that passes, of a continuous flow going from the future toward the past .... Entities have to be made contemporary by moving in step and have to be replaced by other things equally well aligned if time is to become a flow. Modern temporality is the result of a retraining imposed on entities which would pertain to all sorts of times and possess all sorts of ontological statuses without this harsh disciplining. (Latour 1991, 72)

Latour's hypothesis is that moderns constructed such a temporality as a consequence of the purification of subjects and objects: 'The idea of radical revolution is the only solution the moderns have imagined to explain the emergence of the hybrids that their Constitution simultaneously forbids and allows, and in order to avoid another monster: the notion that things themselves have a history' (Latour 1991, 70). Since nature is always already given and all of history only involves social subjects, a history of science becomes an oxymoron: impossible, if it were not for the fact that new entities do seem to appear throughout history. The only solution is to give the history of science a specific logic, separated from ordinary history:

> From now on there will thus be two different histories: one dealing with universal and necessary things that have always been present, lacking any historicity but that of total revolutions or epistemological breaks; the other focusing on the more or less contingent or more or less durable agitation of poor human beings detached from things. (Latour 1991, 71)

But the proliferation of quasi-objects, that resist to be classified as either subject or object, past or present, show that purification was never the full story, since it never had 'any effect on the practice of mediation, a practice that has always mixed up epochs, genres, and ideas as heterogeneous as those of the premoderns' (Latour 1991, 69).

But if we are not modern, what are we then? Are we premodern? No, since that category only functions as shorthand for not-yet-being-modern. Should his position then be classified as 'postmodern' (Latour 1990)? Here Lyotard re-enters the scene:

> Postmodernism is a symptom, not a fresh solution. It lives under the modern Constitution, but it no longer believes in the guarantees the Constitution offers. It senses that something has gone awry in the modern critique, but it is not able to do anything but prolong that critique, though without believing in its foundations (Lyotard 1979). (Latour 1991, 46)

According to Latour, the postmodern position of Lyotard does not escape from the problems posed by the quasi-objects. On the one hand, though postmodernists might contest the modern claims of being completely separated from the premoderns by making pastiches between premodern and modern elements, they nevertheless invoke a new historical break to determine their own point of view. On the other hand, they still 'accept the total division between the material and technological world on the one hand and the linguistic play of speaking subjects on the other – thus forgetting the bottom half of the modern Constitution' (Latour 1991, 61). To illustrate this, Latour gives an extensive quote of an interview with Lyotard in *Le Monde*:

> I simply maintain that there is nothing human about scientific expansion. Perhaps our brain is only the temporary bearer of a process of complexification. It would then be a matter of detaching this process from what has supported it up to now. I am convinced that that is what you people [scientists!] are in the process of doing. Computer science, genetic engineering, physics and astrophysics, astronautics, robotics, these disciplines are already working toward preserving that complexity under conditions of life independent of life on Earth. But I do not see in what respect this is human, if by human we mean collectivities with their cultural traditions, established in a given period in precise locations on this planet. I don't doubt for a second that this 'a-human' process may have some useful fringe benefits for humanity alongside its destructive effects. But this has nothing to do with the emancipation of human beings. (Lyotard 1988a, xxxviii)

For Latour, such a strong opposition between cold objects of science and the meaningful lifeworld of subjects is untenable, since it once again denies the reality of quasi-objects. Similar to Serres, in Latour's philosophy science and technology are never fully purified from myth and culture: 'Where are the

Mouniers of machines, the Levinases of animals, the Ricoeurs of facts' (Latour 1991, 136)?

Postmodernity, just as modernity, does not recognize that the purification of subjects and objects in space and time is never a given, but a practice. Temporality is thus a way of connecting things. What Latour proposes is not so much a criticism of this organizing principle, but instead an alternative. The amodern framework does not start from a linear time, with continuities and discontinuities, but rather uses the metaphor of the 'brewing of history' (Latour 1991, 68) or 'exchangers and brewers of time' (Serres and Latour 1992).[2] Latour uses Serres's image of time, according to whom time follows not so much a line, but a spiral:

> We do have a future and a past, but the future takes the form of a circle expanding in all directions, and the past is not surpassed but revisited, repeated, surrounded, protected, recombined, reinterpreted and reshuffled. Elements that appear remote if we follow the spiral may turn out to be quite nearby if we compare loops. Conversely, elements that are quite contemporary, if we judge by the line, become quite remote if we traverse a spoke. Such a temporality does not oblige us to use the labels 'archaic' or 'advanced', since every cohort of contemporary elements may bring together elements from all times. In such a framework, our actions are recognized at last as polytemporal. (Latour 1991, 75)

Such a spiral metaphor offers room to recognize quasi-objects and the history of natural things (Serres 1989; Latour 1999b): things can have a history, since this history consists in the creation of new links and the abandonment of earlier ones; these connections can be elements that have only recently entered the scene or old elements that have been around for long. In his interview with Latour, Serres gives the example of a car:

> What things are contemporary? Consider a late-model car. It is a disparate aggregate of scientific and technical solutions dating from different periods. One can date it component by component: this part was invented at the turn of the century, another, ten years ago, and Carnot's cycle is almost two hundred years old. Not to mention that the wheel dates back to neolithic times. (Serres and Latour 1992, 45)

We are thus not faced with a human history against a fixed natural décor, which occasionally brings forth radically new entities such as air pumps, electrons or DNA, but rather a spiral that reciprocally defines society and nature, in relation to the quasi-objects they turn around: 'Each loop in the spiral defines a new collective and a new objectivity. The collective in permanent renewal that is organized around things in permanent renewal has never stopped evolving' (Latour 1991, 84–5). Similarly, the specificity of science is defined by the 'extension of the spiral, the scope of the enlistments it will bring about, the ever-increasing lengths to which it goes to recruit these beings' rather than 'some epistemological break that would cut them off forever from their prescientific past' (Latour 1991, 108).

This also sheds a new light on Serres's Great Story, which similarly implies a different temporality. First of all the Great Story does not follow a linear path, but rather is 'bouquet-shaped':

> we would have a narrow conception of it, even a false one, if we thought it to be linear and only directed towards us, as though we were playing the role of the goal and end of all things . . . whereas it explodes and bifurcates in a thousand contingent ways like a tremendous flowering and ends up today at our present of course, here and now, in front of this old farm, in company with my granddaughter, but also at as many different existences as there are galaxies, black holes or bits of star dust in space, living things in the rain forests and oceans, men and women, cultures and languages on this planet. (Serres 2003a, 13)

Secondly, time is structured chaotically, in the sense that it can always unpredictably branch off in totally novel ways. Only retrospectively does it gain the impression of being determined: 'Unpredictable when it advances, it becomes deterministic when one turns around. Like every narrative, this one, the greatest and most truthful of all, unfurls the contingent time of chaos' (Serres 2003a, 13). Finally, there is no clear endpoint. Although we can understand the path, even understand how we got here, we cannot predict where we are going. At best, the contemporary sciences can tell us where not to go, for instance, towards a nuclear winter or a planet completely in the grip of the greenhouse effect.

Lyotard's postmodern fable

The work of Serres and Latour thus challenges that of Lyotard on multiple fronts: one should be sceptical of historical breaks, universalism is not defeated but can be renewed and an updated humanism is possible. But rather than concluding that Lyotard's philosophy is simply incorrect, I will argue that he can respond to these criticisms and even ends up in a position similar to that of Serres and Latour. This is not accidental, since Lyotard was directly inspired by them. For instance, in one of Lyotard's letters from *Le Postmoderne expliqué aux enfants*, he draws heavily on their relational ontology:

> There can be a science of science – and there is – just as there is a science of nature. The same goes for technology: the whole field of STS (science-technology-society) appeared within a decade of the discovery of the subject's immanence in the object it studies and transforms. And vice versa: objects have languages; to know objects you must be able to translate their languages. Intelligence is therefore immanent in things. In these circumstances of the imbrication of subject and object, how could the ideal of mastery persist? It gradually falls out of use in the representations of science made by scientists themselves. Man is perhaps only a very sophisticated node in the general interaction of emanations constituting the universe. (Lyotard 1986, 20–1)

But the main reason to doubt that Lyotard's postmodernism is fundamentally different from the amodernism of Serres and Latour is the fact that one can find Serres's Great Story in Lyotard's work as well. To quote another of Lyotard's letters:

> The new stage is slowly being set. Some of the highlights: the cosmos is the fallout from an explosion; the force of the original shock is still scattering the debris; as they burn, the stars transmute the elements; their time is running out; so is the sun's; the chances of a synthesis of the first algae occurring in the water on earth were minuscule; the human being is even less probable; its cortex is the most complex material organization that we know; the machines this cortex engenders are its extensions; the network these machines will form will be like a second and even more complex cortex. (Lyotard 1986, 84–5)

But as this quote already clarifies, an additional element is brought into play that was not centre stage in Serres's Great Story, namely the death of the sun. It becomes the central element of Lyotard's most explicit version of the Great Story, which he calls the *Postmodern Fable* (*moralité postmoderne*). This fable follows the Great Story of Serres: the creation of the universe, of life, of *homo sapiens* and finally of written history. Lyotard also interprets the victory of liberal democracies as a mere product of a selective mechanism that chooses the most efficient system to work with. It ends, however, with a projection on the future:

> At the time when this story is told, all the research . . . was in fact devoted, directly or indirectly, to test and reshape, or to replace, the so-called 'human' body, so that the brain remains able to function using only the energy resources available in the cosmos. This is how the preparation for the final exodus of the negative-entropic system took place, far away from the Earth. What the Human

and his Brain, or rather the Brain and its Human, could look like when they left the planet forever, the story did not tell. (Lyotard 1986, 85–6)

This element also provides a response to Serres's accusation that this is a metanarrative. Lyotard states that, despite being a story, 'the fable does not have the hallmarks of a modern "meta-narrative"' (Lyotard 1986, 93). First of all, because 'the human species is not the hero of this story. It is a complex form of organized energy. Just like other forms it is undoubtedly transient. Other more complex forms can arise, which will gain the upper hand' (Lyotard 1986, 87). Energy is the protagonist and the fable follows the internal struggle of energy in the form of negentropic forces struggling in an entropic universe.

But this does not mean that energy is the subject, since energy does not have the relevant properties: 'What happens to the energy, their arrangement in systems, their death or survival, the appearance of more differentiated systems, the energy knows nothing about it and *wants* nothing from it. It obeys blind local laws and coincidences' (Lyotard 1986, 47). More broadly, Lyotard argues, the fable lacks the essential properties of historicity. For Lyotard this means at least four things. 'It is first of all a physical story, it is only about energy and about matter as an energy state' (Lyotard 1986, 91). Secondly, the temporality involved is nothing but a diachronic time. 'This time is not a temporality of consciousness which demands that the past and the future, in their absence, nevertheless be held 'present' at the same time as the present' (Lyotard 1986, 91). Thirdly, there is no teleology nor a horizon of emancipation. Ideally something is saved, although there is no guarantee, but this does not lead to a greater understanding, but is merely 'the result of cybernetic feedback with controlled growth' (Lyotard 1986, 92). Finally, therefore, this projected future is no object of hope. There is no ultimate promise that we will end up fine, since it will not be us that survive the death of the sun.

Similar to, yet different from Serres, Lyotard sees this fable as the final blow for human narcissism, after Copernicus, Darwin and Freud. Through this postmodern fable, the reader

> learns that s/he does not have the monopoly of mind, that is of complexification, but that complexification is not inscribed as a destiny in matter, but as possible, and that it takes place, at random, but intelligibly, well before him/herself. S/he learns in particular that his/her own science is in its turn a complexification of matter, in which, so to speak energy itself comes to be reflected, without humans necessarily getting any benefit from this. And that thus s/he must not consider him/herself as an origin or as a result, but as a transformer ensuring, through

techno-science, arts, economic development, cultures and the new memorization they involve, a supplement of complexity in the universe. (Lyotard 1988b, 45)

To the criticism of Latour that there has never been a historical break, Lyotard is able to reply as well. Though Lyotard links postmodernity in *La condition postmoderne* with a specific moment in time, in later works he stresses that both modernity and postmodernity are not historical periods: 'Modernity is not an epoch but a mode (the word's Latin origin) within thought, speech, and sensibility' (Lyotard 1986, 24). Similarly, Lyotard argues:

> Postmodern does not mean recent, but indicates the status of writing, in the broadest sense of thought and action, after the contamination by modernity and after the attempt to recover from it. By the way, modernity is not recent either. It is not even an era. It is another state of writing, in the broad sense. (Lyotard 1986, 89)

Similar to Latour, Lyotard thus sees modernity as a specific way to organize events and phenomena in time, linked for Lyotard with metanarratives and their specific structure: 'It gives modernity its characteristic mode: the *project*, that is, the will directed toward a goal' (Lyotard 1986, 50). Subjects and objects are thus arranged in accordance with that project as either contemporary or obsolete, purified or mixed.

Relational ontology and performativity

Lyotard's position thus comes close to that of Serres and Latour, yet they differ in their evaluation: whereas Serres is ultimately positive, Lyotard is clearly negative. What is at stake in Lyotard's work is an issue common to all relational ontologies, that is ontologies that analyse the world as networks without any given essences or hierarchies: How to make a distinction between good and bad constructions, good and bad relations? *Performativity* is the name for the risk of the complete erasure of any kind of normative qualification of relations. Anything could then be exchanged for anything; anything is up for debate. Nothing is sacred.

From codes to axiomatics

Let us first revisit the problem of performativity. Often, in secondary literature, this issue in Lyotard's oeuvre is simply interpreted as a critique of capitalism (e.g. Williams 1998). But what is at stake for Lyotard is more fundamental, namely a specific logic of organizing relations. Lyotard defines this logic as follows:

> It results from a process of development, where it is not mankind which is the issue, but differentiation. This obeys a simple principle: between two elements, whatever they are, whose relation is given at the start, it is always possible to introduce a third term which will assure a better regulation. *Better* means more reliable, but also of greater capacity. (Lyotard 1988b, 6)

Once again, this is no metanarrative. It does not matter what is optimized, as long as optimization occurs. There is only one criterion by which elements are evaluated and dismissed: 'Improve performances' (Lyotard 1988b, 199). Already in *La condition postmoderne* Lyotard warns for this: 'The speculative hierarchy of learning gives way to an immanent and, as it were, "flat" network of areas of inquiry, the respective frontiers of which are in constant flux' (Lyotard 1979, 39). But its most crucial consequence is found in Lyotard's Postmodern Fable, since, although it has no finality, the logic has a limit, namely the death of the sun:

> The anticipated explosion of this star is the only challenge objectively posed to development. The natural selection of systems is thus no longer of a biological, but of a cosmic order. It is to take up this challenge that all research, whatever its sector of application, is being set up already in the so-called developed countries. The interest of humans is subordinate in this to that of the survival of complexity. (Lyotard 1988b, 7)

This is not an intentional process, but rather a network of relations that develop according to a logic, over which we are losing control. Lacking full control is not necessarily problematic, but what is a problem is that humanity risks to disappear. The logic of performativity is driven to a point where the question becomes 'how to make thought without a body possible' (Lyotard 1988b, 13)? Though humans might currently still be a useful node in it, there is no guarantee that this will remain so in the future. It is in this sense that Lyotard calls this form of science 'inhuman', as in the interview quoted by Latour (1991, 61).

Performativity thus threatens humanity since it is indifferent to the nature of the nodes in the network. Instead performativity makes everything interchangeable with a higher efficiency in mind: 'be operational (that is, commensurable) or disappear' (Lyotard 1979, xxiv). These themes echo Lyotard's earlier work of *L'économie libidinale* (1974), where performativity is interpreted as maximizing libidinal flows. Indeed, one could read Lyotard's book in the line of this argument, but it is perhaps made even clearer in Deleuze and Guattari's book *L'Anti-Œdipe* (1972), which had a similar ambition. To oversimplify, the book contains a history of capitalism in three stages: nomadism, the despotic states and capitalism itself. In nomadic societies relations are controlled by filiation and alliance, for example in terms of kinship and other codes that limit the way entities can interact. This is similar to how for Latour premoderns control their hybrids, as Latour recognizes with a reference to Deleuze (Latour 1991, 117). With the invention of the state, however, the 'coded flows of the primitive machine are now forced into a bottleneck where the despotic machine overcodes them' (Deleuze and Guattari 1972, 199). The existing codes are thus 'overcoded' by a central, hierarchical and often transcendent code, putting the State (and its king, for instance) at the centre as an obligatory passing point.

The way Deleuze and Guattari read this history is in terms of fear: fear of decoding, a fear that without any code all relations become possible and an endless accelerating flux would destroy any meaningful order. This the authors identify with capitalism: 'In a sense, capitalism has haunted all forms of society, but it haunts them as their terrifying nightmare, it is the dread they feel of a flow that would elude their codes' (Deleuze and Guattari 1972, 140). Thus, in the third stage, that of capitalism, we witness a process of decoding (or 'deterritorialization'), where codes are mobilized, changed, uprooted and destroyed in order to optimize flows. In that sense, although it announces the nightmare of complete decoding, capitalism still keeps certain codes at work (a process of 'reterritorialization'), rather than abandoning them all. Capitalism is 'the *relative limit* of every society, in as much as it axiomatizes the decoded flows and reterritorializes the deterritorialized flows' (Deleuze and Guattari 1972, 266). Deleuze and Guattari speak of axiomatization, since what is at stake in these codes is no longer their representational value: they do not aim to meaningfully situate individuals in the world. Rather, working on flows of energy, randomly chosen conventions regulate them like a traffic light or an electrical grid. No relation is sacred and all are organized in function to optimize the negentropic process of order within disorder. The Postmodern Fable recounts the story of this 'nightmare' of performativity.

Should performativity be criticized?

The diagnoses of Lyotard, Deleuze and Guattari are not that foreign to Serres or Latour, mainly because they all share a relational ontology. But whereas for the former this is clearly something negative, for Serres and Latour it seems to be, at first sight, something to be welcomed and even celebrated. We found in Serres and Latour, for instance, a new humanism that seems to thrive on this decoding of relations. Serres's humanism, for instance, entailed the positive appraisal of denser networks, implying more relations with others:

> The more you imprint others onto the self, the more it becomes established as singular – for no one else presents this remarkable colour – but also the more it runs towards a sum as white as the wax we began with. This whiteness can be regarded either as a colour or as the integration of every colour. Pierrot tends towards Harlequin; Harlequin tends towards Pierrot; this double incandescence forms human time. So universal humanity becomes that virginity that was received at birth, achieved at death and whose plastic becoming we recognise in ourselves. We will therefore give it two identity cards, the one white, the other multicoloured. (Serres 2003a, 84)

Latour, in a similar vein, celebrates not only this humanism but an accompanying shift from science to what he calls research. In the traditional model 'science' and 'society' are purified, research implies a model where they 'are now entangled to the point where they cannot be separated any longer' (Latour 1998, 208). It is quite easy to read the modern work of purification, separating science and society, as a form of coding, that is being dismantled, with Latour celebrating this process:

> If Science thrived by behaving as if it were totally disconnected from the collective, Research is best seen as a *collective experimentation* about what humans and nonhumans together are able to swallow or to withstand. It seems to me that the second model is wiser than the former. (Latour 1999b, 20)

Both Serres and Latour are familiar with the claims of Lyotard and Deleuze. Serres, for example, referred to Deleuze as his 'best friend. I admired him. I loved him' (Serres 1995). Latour stated in an interview: 'I have read Deleuze very carefully and have been more influenced by his work than by Foucault or Lyotard' (Latour and Crawford 1993, 263). Nonetheless, they dissociate themselves from Lyotard and Deleuze's fears:

> How could the *anthropos* be threatened by machines? It has made them, it has put itself into them, it has divided up its own members among their members, it has built its own body with them. How could it be threatened by objects? They have all been quasi-subjects circulating within the collective they traced. It is made of them as much as they are made of it. It has defined itself by multiplying things. (Latour 1991, 137–8)

A similar criticism is found in Serres: 'The entirety of our devices, themselves having set sail from our body, has the effect of augmenting, along the same line, this totipotency How would technology dehumanize humanity since it extrapolates the best corporal characteristic from it' (Serres 2001, 121)?

At first sight these rebuttals seem strange, since they rely on the modern idea that we have full control over that which we construct. An alternative reading is that they aim to problematize the opposite position, the belief that we are doomed to be the slaves of technology:

> Those who agree that capitalism is really deterritorialized, technology really sleek, discourse really empty, society a real simulacrum, and science totally and fully unhuman, abandon the field without having fought. They leave to the enemy a territory much larger than the one they had. It is a sort of intellectual Munich. I might be too severe; I might be biased by our recent French political history. Postmodern theorists are useful, like salt added to the academy. A pinch of Lyotard, a pinch of Baudrillard might be good, but a whole meal of salt? (Latour and Crawford 1993, 254)

Although I believe Latour and Serres are correct in this criticism, Lyotard nor Deleuze endorse this view. One can accept that technology and humanity are not fundamentally opposed, while accepting that concrete, historical networks form a threat to the relations, networks or territories we hold dear. When Lyotard speaks about the inhuman aspect of science he is not stating that humans have no control whatsoever, but rather that the existing infrastructures and networks of science and technology threaten the network known as the 'human'. It is thus possible to acknowledge that the human has no fixed essence, but still argue that within its current form there are certain relations that are not up for debate. If one only focuses on how all relations can be renegotiated, without asking the question whether they all should, one becomes an ally of this logic of performativity. There is thus a fallacy at play, mixing up the refutation of the idea that there is no *a priori* stability with the refutation that there are also no local and temporal forms of stabilities, which can still be cherished and defended (see final chapter).

In fact, one can find this other side as well, in the work of Latour and of Serres. At the end of *Nous n'avons jamais été modernes* Latour's message is that we should not give up all forms of purification, and he even actively pleads 'to keep the moderns' major innovation: the separability of a nature that no one has constructed – transcendence – and the freedom of manoeuvre of a society that is of our own making – immanence' (Latour 1991, 140). But what he wants to see changed is that the process of purification should become explicit and open for debate: 'We want the meticulous sorting of quasi-objects to become possible – no longer unofficially and under the table, but officially and in broad daylight' (Latour 1991, 142). In that sense, his argument is not one in favour of 'anything goes', but rather that by acknowledging that all relations are constructed, we do this in an explicit manner.

We find similar elements in Serres's work, for instance in his reflections on agriculture, whose disappearance forces us to face the question of how to shape our collective anew. Let me give three extra examples. First of all, the disappearance of place in our society. Whereas all our addresses used to be grounded in the (agricultural) territories that shaped our lives, recent technological innovations have challenged this. 'For the first time, the portable phone and laptop have liberated addresses from places' (Serres 2001, 180). By de- and reterritorializing countless relations, destroying the old networks that situated us, these new technologies open up the question where to situate ourselves:

> Light and instantaneous, the new communications therefore don't merely change the form of the old lattices or the density of their meshing, but transform habitable space itself, the face of the land, tomorrow deserted perhaps by plant roots and animal niches. At the very moment when we are talking about nothing but networks, there are no more networks – merely this new expanse without distance measurement, without physical nodes or crossroads where moving things circulate. How then are we going to inhabit it? (Serres 2001, 151)

A second example is what Serres calls 'world-objects' (*objets-mondes*), a notion he already introduced in his *Hèrmes* series, referring to satellites or nuclear power (Serres 1974a, 101). He picks it up again in *Hominiscence*:

> If I call a constructed tool for which at least one of its dimensions attains the extension of one of the world's dimensions a world-object, the laptop and the mobile telephone, in fact, attain global space and real time: therefore these are world-objects. (Serres 2001, 193)

These world-objects radically destabilize our current relations, raising the problem of ecology, as we will see in the final chapter: 'Less and less objects, more and more world, world-objects lead us to a world which isn't an object like the objects of the world' (Serres 2001, 141). A third example is money. By making everything interchangeable it disrupts existing relations in novel and problematic ways:

> White money is worth everything, can do everything, rules over everyone, omnipotent; the administrator who manages money in our societies obtains a hyperpower there because he parasitizes it. Money and administration devour

the entire social bond. . . . In the West, money has invaded everything, destroyed everything, by taking everything up into its measure. Traditional societies, consequently, try to defend the ties they fear to see destroyed or undone; they try to bind themselves again. But do we know what bond gathers us together? (Serres 2003a, 55)

There is clearly a normative concern behind these issues, concerned with the question: Through which relations should we situate ourselves? Serres even links this with the task of metaphysics: whereas humanity, and the process of hominiscence, is characterized by a despecialization of the human, in the sense that he becomes totipotent again, there is a limit to this process:

> What is metaphysics? It describes the minimal thresholds of our despecializations, whether corporal or externalized. The thresholds of whiteness, abstraction, symbol, the low limits below which we cannot plunge without dying. What purpose does it serve? To watch over these dangerous critical points. (Serres 2003a, 66)

Conclusion

This chapter has examined Serres's diagnosis of modernity, and its conception of temporality and history. We saw, in a confrontation with Lyotard, that Serres, Latour and Lyotard actually agree on many points, mainly due to their shared relational ontology. The challenge of any relational ontology is thus the following: How, if all relations are principally open to be revised and reconstructed, to decide which relations to foster and which to abandon and avoid? According to which principles should we use this totipotency of the human? And who should have a say in this process of negotiation?

These questions will be central to the next three chapters. First, in Chapter 6, we will look to the negative side of the story: Which relations are to be avoided? Chapters 7 and 8 then explore Serres's alternative, embodied by themes from religion and ecology.

6

Thanatocracy and the anthropology of science

Introduction

We have seen how Serres starts from a relational ontology (Chapter 4), raising the question of which relations we should cultivate (Chapter 5). This chapter will explore Serres's *anthropology of science*, announced in his text on Bachelard (see Chapter 3):

> Everyone is shocked that knowledge is no longer wisdom: it never has been since its first formation, since the criminal act was its act of birth, in the shade of Jupiter's flamines and of the legionaries of Mars. Knowledge is allied with power, it is power in its very essence, not only since empires have recognized and stolen its might, but since it has established itself as knowledge at the loci of strategy, of conquest and of the empire. *And now, under penalty of death, we are forced to outline a more archaic prehistory than that of Bachelard, to purify the sources of science poisoned, from the beginning, by terror.* It is no longer a question of the fine list of capital sins, but of our collective survival in the face of capital punishment. (Emphasis in original) (Serres 1970, 45)

With 'capital punishment' Serres is referring to our potential collective death due to the atomic bomb, a central topic of Serres's text 'Betrayal: Thanatocracy' (Serres 1974a), which we will take as our starting point. But this violence is not so much a product of recent developments in science – such as nuclear weapons – but rather an expression of a longer, archaic history of science. In the first part of this book, we saw how science implied violence because it purifies: it reduces the multiplicity of phenomena to a limited set of categories. But there are two other types of violence produced by science as well. First of all, there are the violent effects of new technologies and scientific insights which result in what Serres calls a 'thanatocracy', leading to the possible destruction of our collective world. Secondly, in the form of the academic practice of critique, there is also violence against other subjects who stand accused and surveilled. These two forms of violence will be the topic of this chapter.

To map this archaic history of violence, Serres relies on two authors: Georges Dumézil and René Girard. There is hardly any secondary literature on their relation to Serres (though see Sayes 2012). I will therefore start with Dumézil's trifunctional hypothesis on Indo-European mythology, exemplified by the Roman gods of Jupiter, Mars and Quirinus. Then, I will focus on Girard's scapegoat hypothesis and theory of mimetic desire. Girard has also been picked up by other authors, associated with the Group of Ten (*Groupe des Dix*), who has translated Girardian themes to other scientific domains. This is not what makes Serres's take unique. In the final section therefore I explore how his approach is innovative in (a) his combination of Girard with Dumézil; (b) his focus on the role of quasi-objects as scapegoats, linking it to Quirinus, Mars and Jupiter; (c) as well as in his application of their insights, not only to the history of science and technology but also to the theme of critique: criticism and the social sciences can be understood as an extension of, rather than the end of mythical violence. Unearthing these hidden histories of violence in science and critique constitutes the main object of Serres's anthropology of science.

The sound of Hiroshima

Serres often describes himself as 'a child of Hiroshima' (Serres 2014, 32), one of the most significant historical events in history. Serres was, however, struck by how his fellow philosophers of science ignored this event. We already saw (in Chapter 1) how Serres dismissed Bachelard's new scientific spirit, partly because it never 'heard the sound of Hiroshima' (Serres and Latour 1992, 11). In a later interview, Serres elaborated:

> I remember that for my epistemology teachers, Gaston Bachelard or Georges Canguilhem, the question did not even arise. For example, Bachelard published *L'Activité rationaliste de la physique contemporaine* in 1951. In it, there is no word about the relationship between science and society. Quantum mechanics lead to the atomic bomb but there is not the slightest allusion to Hiroshima, neither there nor in any subsequent work! (Serres 2014, 160)

Since epistemology has lost its traditional task (see Chapter 1), Serres suggests an alternative, namely an *anthropology of science*. This refers to a deeper, more archaic history of science, parallel to traditional history of science (Serres 1987, 6). The object of this anthropology of science is the often unrecognized violence at work within science. Hiroshima embodies not so much the start of this violence, but a threshold where it became too omnipresent to ignore. 'A philosopher who listens to science hears today, among technical information, a word of death. The death of our world and of mankind' (Serres 1974a, 72).

This violence is thematized in an early text, 'Betrayal: Thanatocracy' (Serres 1974a), in which Serres argues that Hiroshima embodies several crucial societal shifts. First of all, it leads to a new idea of death. Besides the traditional conceptions of death as a dying individual or a collapsing civilization, 'there is a third [kind of death] that appears on the horizon of Hiroshima, and it is the death of the species. It is the death of mankind and the Earth' (Serres 2014, 166). For Serres, humanity has become 'collectively suicidal' (Serres 1974a, 78).

Secondly, this entails a shift of focus for philosophy of science, which should concern itself with our *limits*. This does not concern the transcendental limits of reason or our anthropological finiteness, but our physical limits: the threshold of what the world can bear. Once purely speculative, we have started to experience these limits through our actions. 'It is not theoretical, it can be extrapolated from our interventions. Therefore, any future ethics [*déontologie*] can only be thought of with reference to this limit' (Serres 1974a, 97). Nuclear weapons refer to those limits by transgressing them: were we to use these weapons, we would destroy not only our enemy but also ourselves.

Thirdly, for Serres nuclear weapons are a type of 'world-objects' [*objets-mondes*]: 'Objects with the dimensions of the world, in the precise sense of the dimensional equations: for space (ballistic missiles), for the speed of rotation (geostationary satellites), for time (the lifetime of radioactive waste), for energy and heat' (Serres 1974a, 101). Because of the stakes, science has become intrinsically irrational through these world-objects. To Serres it is not a question of whether someone might misuse nuclear weapons, but the fact that they even exist. 'I am not saying: there are dangerous lunatics in power – and one would suffice – I am saying: there are only dangerous lunatics in power. All are playing the same game, and are hiding from humanity that they are planning its death' (Serres 1974a, 73–4).

Finally, Serres also uses the notion of death in a second sense. This concerns death as *inertia*: the history of science loses all novelty and unpredictability by being enslaved to a preset goal, such as war. Science is dead because it became inert: without movement, change, novelty. 'There is no longer really a history of science because it is now overdetermined in its progress. Science is out of history. It has entered a post-historical era. It is, through and through, invaded by the death instinct' (Serres 1974a, 76–7).

In sum, science has become organized around the aim of our collective destruction. This is what Serres calls a thanatocracy: 'The government of death' (Serres 1974a, 80). However, Serres's argument is not that science only became linked with violence after the Second World War. Hiroshima is not a real historical breaking point, but a tipping point where the violence of science enters a new stage. Violence is not only linked to contemporary issues, such as nuclear weapons, but has a longer history. 'We will soon see that you have to go further back, a century first, then two millennia at least, all of history in the end, to understand it' (Serres 1974a, 77). For Serres, it is hard to find a point in history where science is free from any violence: 'Reason is sick to death the instant it emerges. Everything was in place from there, from the Greek miracle, this immense historical catastrophe where logos transpires destruction and homicide. Reason is genocidal by birth' (Serres 1974a, 84–5).

In this context, it becomes clear that Serres's many references to the figures of Mars, Jupiter and Quirinus are not accidental. Hiroshima, for Serres, embodies the moment when science 'is not under the control of Jupiter or Quirinus, priests or producers, but in the hands of the soldiery' (Serres 1974a, 86). Serres borrows these figures (Mars, Jupiter, Quirinus) from Georges Dumézil.

Jupiter, Mars and Quirinus

It is hard to put a label on Georges Dumézil (1898–1986), who could either be described as a French mythographer, a historian of religion or an anthropologist. He was fascinated by languages – he mastered over thirty of them – and their histories. But controversially, he claimed it was possible to link language and mythology with the social and political structure of the societies in which they circulated. Most famously, Dumézil claimed that Indo-European languages were structured around three ideological functions: a juridical-religious, a military and an economic-agricultural one, linked to the three foundational Roman Gods: Jupiter, Mars and Quirinus respectively (Dumézil 1938).

Dumézil's tripartite hypothesis

The idea that a number of European and Indian languages shared a common ancestor – the so-called Proto-Indo-European language – dates back to the end of the eighteenth century. In its essence, it is a claim about language. Most scholars agree that Proto-Indo-European language existed, but the main point of controversy concerns the specifics of its history. Until the Second World War, it was a popular hypothesis that the origin lies with Germanic Aryans (hence Indo-German as a term). In contrast, the now popular Kurgan hypothesis advocated that the Proto-Indo-European language should be traced back to the Kurgan culture in the Pontic steppe, north of the Black Sea. Alternative theories, such as the Anatolian hypothesis and the Armenian hypothesis point at Neolithic Anatolia and Armenia as possible points of origin. The second point of controversy concerns the timing and mode of this original migration. The Kurgan hypothesis situates it at 4000 BCE, while the Anatolian hypothesis traces the origin to around 8000 BCE. They also disagree on the nature of the migration process: whereas the former suggests a migration through military means, due to new military technologies such as the horse, the latter points to peaceful innovations in agriculture to explain the migration.

Dumézil was not so much concerned with questions about the origin of Indo-European languages, but with their structure. He was interested in possible common social and cultural characteristics of Indo-European societies. It is from this perspective that Dumézil (1938) proposed that Indo-European societies are characterized by a tripartite system. There was first of all a *sovereign function,* embodying both sacred and secular leadership. Secondly, there was a *warrior function,* linked to the defence of society. Finally, there was the *economic function,* linked to health and sustenance, agriculture and commerce.

This tripartite ideology, present in the Proto-Indo-European society, spread across the whole domain where Indo-European languages are spoken. Dumézil's methodological claim was that this tripartite ideology could be discovered in most of the early mythical and epical literature in Indo-European languages. Dumézil linked the first function, that of sacral and earthly sovereignty, with pairs of gods, such as the Vedic figures Mitra–Varuna, the Norse gods Tÿr and Odin and the Roman divinities Dius Fidius and Jupiter. The second function, linked to war, was found in the Vedic deity Indra, the Norse god Thor and the Roman god Mars. Finally, the third function, linked with fertility and wealth, was linked to pairs of divinities as well, such as the Vedic Asvins, the Norse Njoror and Freyr or the Greek *Dioscuri.* In the case of Rome, Dumézil linked it to Quirinus.

To what extent Dumézil's system is applicable to all Indo-European societies is an object of controversy. He himself linked it most successfully to Post-Vedic Indian society, mainly due to its rigorous division in four castes (*varnas*), three Aryan castes separated from the fourth. These three castes were linked by Dumézil with the three functions: 'the brahmans, the priests who study the holy scriptures and perform the sacrifices; the kshatriyas (or râjanyas), the warriors who defend the people; and the vaishyas, the producers of material goods' (Belier 1991, 10).

Most famous is Dumézil's application of his hypothesis to the Roman tripartite of Jupiter, Mars and Quirinus. He would also link it to the three Romulan tribes of Ramnes, Luceres, and Tities, respectively the three castes of priests, warriors and producers (Dumézil 1941). Similarly, the first kings of Rome embody the characteristics of the tripartite system. Like the Vedic gods Varuna and Mitra, Romulus and Numa Pompilius can be equated with the double first function, while Tullus Hostilius can be linked with the second, warlike function. The third function is less clear, though Dumézil associates it with the Sabines, who were linked to luxury and voluptuousness.

The case of Ancient Greece resisted Dumézil's hypothesis, which he labelled his '*grande mystère*' (Dumézil 1958, 91). He partly tried to explain away the anomaly by arguing that the specific novel political organization of the polis and influences of the Middle East erased the Indo-European tripartite ideology. But there are some elements that could be linked to the tripartite system, such as the judgement of Paris in Homer's *Iliad*, where the Trojan prince must pick between three gods, Hera, Athena and Aphrodite, each offering him a gift: world dominance, military skill and earthly pleasure. Paris picks Aphrodite and by that choice initiates Troy's downfall (Dumézil 1953). Similarly, one could read Plato's model of the ideal state as a tripartite one, with a class of rulers, warriors and workers (Dumézil 1941, 257).

Impact and criticism

Dumézil's theory has proven to be quite influential. Besides students within his field, the tripartite hypothesis was taken up by scholars such as Roger Caillois, Jean-Pierre Vernant and Claude Lévi-Strauss. Georges Duby famously mapped the tripartite social system of organization within the history of Medieval Europe, dividing up society into priests, kings and peasants (Duby 1978, 5). Notorious is also the connection between Dumézil and Foucault. Foucault later claimed he was inspired in his early work by Dumézil 'through his idea of structure. Just as Dumézil does with myths, I attempted to discover the structured forms of experience whose pattern can be found, again and again, with modifications, at different levels' (1961, 8). Institutionally, Dumézil would play a crucial role in Foucault's appointment in Sweden in the 1950s and his election to the *Collège de France* (Eribon 1994, 35–37, 105–83). Dumézil himself was elected to the

Collège de France in 1949, creating a chair for Indo-European civilizations. Didier Eribon, who wrote influential biographical works on Foucault, did the same for Dumézil (Dumézil 1987; Eribon 1992). It was Foucault who introduced Serres to Dumézil (Serres and Latour 1992, 36).

Dumézil himself was not always clear on the scope of his argument and later softened his claims (e.g. Dumézil 1968, 15). Hence, Dumézil had many critics. A first group contested his interpretation of the Indo-European sources or the conclusions drawn from them (e.g. Schlerath 1995, 1996). Arnaldo Momigliano, for instance, claimed that '[t]rifunctionality is altogether alien to the Roman pantheon' (Momigliano 1984, 322). A second group of critics argued that Dumézil's theory was ideologically contaminated by Nazism (see Geroulanos and Philips 2018). Dumézil indeed wrote, from 1933 to 1935, under the *nom de plume* Georges Marcenay for *Le Jour*, an extreme right newspaper, and was involved in circles close to Charles Maurras's *Action Française*. However, Eribon came to his defence, arguing that, though Dumézil was a right-wing conservative, he had no sympathy for Nazism (Eribon 1992, 123-7). The current consensus is that Dumézil had sympathies towards Mussolini's fascism, though not to Nazism.

Serres and Dumézil

Serres regularly refers to Dumézil, especially in *Genèse* (1982a) and *Atlas* (1994), but added a number of features. First of all, Serres reinterprets the three functions as three types of relationships or quasi-objects: that is three ways in which certain objects can structure the social relations surrounding them. The first function organizes the social through a religious fetish or a king, a quasi-object worshipped by all around them; the second function structures the social by struggle or conflict, through a weapon or an army; the third function uses money or commodities to translate social relations into economic ones (fetishes in the Marxist sense). 'Jupiter makes us believe in the all-religious; Mars imposes the all-military; Quirinus today makes us think it all boils down to the economy' (Serres 2008a, 112).

Secondly, Serres is also interested in the connections and alliances between these three functions:

> Georges Dumézil rarely dwells on the possible relationships between the three deities: the vestal Tarpeia, for example, belongs to the third, since her dead body is covered with gold and jewelry, but Dumézil's book ignores her stoning, a strange omission for a fate with unforgettable final drama. As this lynching is

the height of violence, should we not consider the relations between Quirinus and Mars? (Serres 1994, 217–18)

Thirdly, Serres's take is more normative. These three dominant types of quasi-objects are problematic since they make us forget the world and the object. 'No, the world is not reduced to fetishes, nor to the stakes of battle, nor to the fetishism of money and commodities' (Serres 2008a, 113). We will see how they are connected with physical forms of violence as well. Therefore Serres sometimes offers a reading of history as an emancipation from these three functions, in search of a fourth one:

> The Age of Enlightenment attempted to free us from Jupiter, I mean from the grip of the diviner. Does it succeed? After the Second World War, a few men of talent and goodwill invented a Europe without borders to try, in turn, to free their nations from the grip of Mars. Will they succeed? Do we now have to free ourselves from the clashes triggered by Quirinus's grip, I mean by production, labor, depletion of resources, trade, economy, the volatile circulation of goods and signs? What New Lights will liberate humanity from these three false gods? (Serres 2008b, 93)

Science often plays a role in Serres's articulation of this alternative, albeit in an ambiguous way. Sometimes science is seen as studying an object which 'is not a fetish, not a stake, not merchandise' (Serres 1982a, 90). At other places, Serres recognizes that science has forged alliances with these three functions: 'Its objects have become fetishes to be worshiped, prizes and competitive stakes, and desirable merchandise. Science is returning to the most archaic of societies' (Serres 1982a, 90–1).

Finally, Serres combines Dumézil's model with the work of René Girard, in order to historicize Dumézil's scheme and to make it more dynamic. It is well known that Serres labelled Girard 'the Darwin of the human sciences' (Girard and Serres 2007, 63). What is often forgotten is that he labelled Dumézil the Linnaeus of the social sciences, requiring a Darwin (viz. Girard) to give his model an evolutionary engine:

> Enlightening and verifiable . . . the tripartite division proposes, therefore, names or figures for classes, but without giving reasons for it; orders species or genera, but without revealing the principle of classification: there you have it, *mutatis mutandis*, a systematics and a taxonomy without an evolutionary engine, or Linnaeus without Darwin. The necessary energy, producing disorder, crises, explosions, movements and various orders, is offered by violence itself, inexhaustibly, according to René Girard. With regard to human groups, the latter

would therefore be to Darwin what Georges Dumézil is to Linnaeus, because he proposes a dynamic, shows an evolution and gives a universal explanation. (Serres 1994, 219–20)

By turning to Girard, we thus reach 'the point where Linnaeus, the classifier, can be reconsidered in the manner of Darwin, the evolutionist' (Serres 1982a, 81).

René Girard and mimetic violence

René Girard (1923–2015) was a French anthropologist and literary critic who became known for his theories about mimetic desire and the scapegoat mechanism. Since there already exists extensive literature on Girard (see Haven 2018), I will only highlight some of his basic ideas and how they relate to Serres's thought and biography.

Serres met Girard in Buffalo in 1975 and they became lifelong friends. In the 1980s, they would both be appointed at Stanford University. Girard's thought is implicitly present throughout Serres's entire oeuvre, but especially in *Genèse* (1982a), *Rome* (1983b), *Atlas* (1994) and *La Guerre Mondiale* (2008a). They all deal, in one way or another, with the creation of social order out of disorder. In turn, Serres would often offer Girard new material from mythology for his theory, for example in Girard's interpretation of Pliny in *The Scapegoat* (Girard 1982, 92–3). Both authors would contribute to special issues and volumes dedicated to each other's work (Serres 1982b; Girard 2005) and Serres would sponsor Girard's election to the prestigious *Académie Française* in 2005 (Girard and Serres 2007).[1]

Mimesis and violence

The two main concepts in Girard's work are *mimesis* and the *scapegoat mechanism*. Let us start with the first. The claim that desire is mimetic is typically read as an empirical claim. However, it is better understood as a philosophical claim, whose status perhaps comes closer to a principle such as causality. One accepts this principle because one believes that it can help to better describe a number of phenomena, whereas the principle itself cannot be empirically tested. At best, it can be pragmatically assessed: it is acceptable if it proves useful to get a grip on the complexity of the world.

Mimesis, within Girard's framework, is thus primarily a universal phenomenon: not only practiced by all, but also a characteristic of all desires. It is important, however, to distinguish between desires and appetites. 'Appetites for things like food or sex – which aren't necessarily connected with desire – are biologically grounded' (Girard 2008, 56). In the case of desires, the subject mirrors the behaviour of the other person, focused on the object of their desire: 'desire itself is essentially mimetic, directed toward an object desired by the model' (Girard 1972, 146).

Although a child imitating a parent is mimetic, Girard is particularly interested in mimesis of possession: one wants what the other has. This then leads to a mimetic triangle with the subject, the other and the object of desire, in which the other also serves as the mediation between you and the object of your desire. The other is both a mediator – since one desires the same – and an obstacle – because only one person can possess the object of desire. According to Girard, there are two types of mediators: external and internal ones. External mediators are remote idols, the ones we encounter in books or on television, without living in the same environment. These cases are relatively benign. Internal mediators, however, refer to cases in which the idol lives in the same environment. These cases, according to Girard, typically lead, not just to violence, but to *mimetic violence*: I desire to hurt the other because she has hurt me.

Eventually, the object tends to disappear and the rivals become doubles, copying each other's behaviour. As a consequence of this mimetic violence, others are getting involved as well, through their own mimetic desires: 'if two persons are fighting over the same object, then this object seems more valuable to bystanders' (Girard 2008, 64). This mimetic crisis is linked not so much to fundamental differences dividing social groups, but rather to the problem of homogenization: the disappearance of all differences, making all subjects into doubles and rivals of one another.

Serres will take over this elementary principle, that violence is always present and expanding. He even compares it to a physical constant: 'violence obeys, in groups, constants similar to those of energy. Just as mechanics and thermodynamics base their exact truths on a stable amount of force in the Universe, does politics rest on a permanent volume of violence in communities' (Serres 2001, 304)?

The scapegoat mechanism

Mimetic desire thus leads to mimetic violence, which escalates. Were it to escalate indefinitely, no order could exist. But there are functional societies, so there must be a way out: the scapegoat mechanism. Due to contingent factors, one individual or group is singled out as the source of all violence, drawing mimetically all violence towards it, and through its eradication the social order is restored. Or in Girard's words:

> When a community succeeds in convincing itself that one alone of its number is responsible for the violent mimesis besetting it . . . then the belief becomes a reality, for there will no longer exist elsewhere in the community a form of violence to be followed or opposed, which is to say, imitated and propagated. In destroying the surrogate victim, men believe that they are ridding themselves of some present ill. And indeed they are, for they are effectively doing away with those forms of violence that beguile the imagination and provoke emulation. (Girard 1972, 81–2)

The scapegoat is thus a focal point of all mimetic violence; it absorbs it and pacifies society, at least momentarily. This gives the scapegoat a particular paradoxical nature as it is being deemed both the cause and the end of violence. A typical characteristic, therefore, is that, once destroyed or chased away, 'the scapegoat becomes divinized in the archaic sense, that is, the all-powerful, Almighty both for good and for bad simultaneously' (Girard 2008, 66).

Girard reads ritual taboos and prohibitions as attempts to repeat and uphold this original sacrificial crisis and its solution. According to Girard, many cultural taboos are therefore linked with violence, blood or moments of indifferentiation: 'Twins are impure in the same way that a warrior steeped in carnage is impure, or an incestuous couple, or a menstruating woman' (Girard 1972, 58). In a similar light, Girard interprets animal sacrifice – where the animal becomes the substitute of the human sacrifice – and the countless ambiguous myths about kings, being both worshipped and sacrificed, as analogous translations of the original scapegoat act (Girard 1972, 11–12).

Here we find a second element taken over by Serres, a paradox: 'Only violence can put an end to violence, and that is why violence is self-propagating. Everyone wants to strike the last blow, and reprisal can thus follow reprisal without any true conclusion ever being reached' (Girard 1972, 26). The way out of this loop is found in making the source of violence somewhere transcendent, seeing 'the process not as something emanating from within themselves, but as a necessity imposed from without, a divine decree whose least infraction calls down terrible punishment' (Girard 1972, 14).

A final, controversial part of Girard's theory is the role of Christianity. Unlike other religions, Christianity exposes the scapegoat mechanism and explicitly takes the side of the victim. For Girard, Christianity thus centres on a fundamental anthropological insight, namely that, once exposed, the social role of the scapegoat mechanism loses its power. Think of the story of Cain and Abel, whose murder is condemned, or the story of Joseph in Genesis, where first Joseph is scapegoated by his brothers, but in turn forgives them. Similarly, 'Job was innocent. The suffering servant in Isaiah is innocent. Jesus is innocent' (Girard 2011, 232).

Girard and Serres regularly invoke the phrase of Simone Weil that 'before presenting a "theory of God", a theology, the Gospels present a "theory of man", an anthropology' (quoted in Girard 2001, 40). Serres will speak of a *metanomics* (*métanomique*): similarly to how metaphysics reflects on the foundations of physics, metanomics reflects on the basis of the human sciences. 'Christianity precisely plays, for the so-called soft sciences, the role of this absent metanomics' (Serres 2008a, 157).

Crossing the Northwest Passage

Serres will use Dumézil and Girard to develop themes that are applicable, not only to the human sciences but to the natural sciences. Through Dumézil and Girard, Serres thus aims to cross the 'Northwest Passage' (Serres 1980b) between the humanities and the sciences, to become an 'instructed third' [*tiers instruit*] (Serres 1991a). But others have also translated Girard's work to the field of science and technology. This does not single out Serres's unique take on the matter. We will look at the Group of Ten (*Groupe des Dix*) of which Serres was a member and in particular the work of Jean-Pierre Dupuy. Only then can we, in the next section, fully realize where Serres's innovations lie.

Groupe Des Dix

Serres was not the only one who brought Girard into contact with science and technology. In fact, between 1968 and 1976 there was a group of French thinkers who had similar ambitions: a group of French thinkers who tried to link science and politics in a new way, the Group of Ten (*Groupe des Dix*), of which Serres was a member (see Chamak 1997). Initially the group consisted of ten members, but over the years others joined. Serres, for example, would join in 1973, along with the biologist Odette Thibault and the politician Michel Rocard.

The Group of Ten originated in 1968, at a conference of *Objectif 72*, a socialist movement founded by the economist Robert Buron. Buron was impressed by the contributions of the neurobiologist Henri Laborit and the philosopher Edgar Morin. Together with the physician Jacques Robin they formed the first four members, though soon it became a group of ten. In addition to the four founders, the group consisted of Gérard Rosenthal (lawyer), Jack Baillet (psychoanalyst), Jacques Sauvan (physician), Jean-François Boissel (computer scientist), Françoise Coblence (philosopher), Bernard Weber (biologist) and Alain Laurent (sociologist). In 1969 three other members joined: the economist René Passet and the politicians David Rousset and Jacques Piette. The biotechnologist and futurologist Joël de Rosnay joined in 1971, while Henri Atlan (biophysicist), Jacques Attali (economist) and André Leroi-Gourhan (paleoanthropologist) joined the group a year later. Serres was one of the last entrants in 1973.

Though mirroring the Macy Conferences in the United States, the Group of Ten was much more politically inspired. Their goal was to find a scientific basis for politics. This went beyond providing relevant facts for policy. Social processes such as elections or economic cycles had to be rethought in terms of complexity and self-organization. More generally, the group was fascinated with the self-organization of complex systems in nature, such as the dissipative structures of Ilya Prigogine (Prigogine and Stengers 1979). Order comes about spontaneously in all sorts of ways, from coffee grounds that sink to the bottom to whirlpools that follow recurring patterns. The group dreamed of applying these models to society, especially in the light of May 68, when the students called for self-organization at university and in politics. Theories such as the scapegoat mechanism of Girard fitted perfectly into these ambitions. The ambition was to write a manifesto, but this failed as it soon turned out that the underlying differences between the members were too great.

Robert Buron's death in 1973 was a turning point. The publication of their *Cahiers des 10* stopped and the hope of finding common ground disappeared. From then on, meetings revolved around discussing the members' new book manuscripts, such as Joël de Rosnay's *Le macroscope* (1975) or Jacques Attali's *La parole et l'outil* (1975). Over the years, the group invited prominent French scientists, such as Jacques Monod and François Jacob (see Chapter 2). In 1972 they also met the Club of Rome. Though agreeing on many points, Jacques Robin later reported how 'the members of the Club of Rome were amazed at the importance we attached to resisting a technocratic view of the world's problems. We were interested in problems of meaning, democracy and questions related to living beings.' The Club of Rome, however, 'talked about politics, but it was a type of politics which left the economic system intact' (quoted in Chamak 1997, 54).

Despite its ambitions, the Group of Ten failed to translate their knowledge into policy impact. Outside the group, there seemed little interest in their message. In a recent interview, Morin disappointedly looked back on it: 'No, we had no influence. Occasionally we had with Robin the ambition to preach to the socialists, to explain to them a thing or two about complexity, but the party was not interested at all' (quoted in Morin, Passet, Vivien and Dicks 2019, 229). The impact of the group was limited to a number of influential publications of its members, ranging from Morin's *La méthode* (1977), Rosnay's *Le macroscope* (1975), Atlan's *Entre le cristal et la fumée* (1979) to Passet's *L'économie et le vivant* (1979). But also the work of Michel Serres, who through these meetings, befriended many of these like-minded intellectuals who shaped his thinking on social order. The uniqueness of Serres's view, therefore, is not that he links questions concerning self-organization and complexity with social phenomena. In order to unearth what is unique, let us have a closer look at one of these figures Serres encountered through this group, Jean-Pierre Dupuy.

The Scapegoat As Quasi-Object

Though not a member of the Group of Ten, Dupuy was invited to one of their final sessions to present his own work on processes of auto-organization and autonomous systems. Through Ivan Illich (see later), Dupuy befriended Heinz von Foerster, the secretary of the Macy conferences, who worked on the concept of auto-organization and the principle of 'order out of noise'. Through von Foerster, Dupuy came into contact with Francisco Varela, who promoted the idea of autopoiesis in biology, and Henri Atlan. But also Serres is a regular reference in his work. On multiple occasions, Dupuy would endorse Serres's ideal of the instructed third (Dupuy 1982a, 20–1; 2013, 47–8), because both felt it to be an appropriate response to the rise of new threatening technologies: 'We are the first to experience the emergence of humanity as a quasi-subject; the first to grasp that it is fated to destroy itself unless avoiding this destiny is agreed to be an absolute necessity' (Dupuy 2013, 62).

Dupuy also often stressed the parallel between Serres and Girard, for instance in how Serres mobilizes the scapegoat mechanism in his analysis of Jean-Jacques Rousseau. In *Le Parasite* (1980a) Serres contrasts Rousseau's theoretical understanding of the emergence of a social contract with his later personal experiences of persecution:

General will is rare and perhaps only theoretical. General hatred is frequent and is part of the practical world. . . . Not only does he see the formation of a social pact from the outside, not only does he notice the formation of a general will, but he also observes, through the darkness, that it is formed only through animosity, that it is formed only because he is its victim. . . . Union is produced through expulsion. And he is the one who is expelled. (Serres 1980a, 118–19)

Dupuy would meet Girard in the United States, and together they organized an interdisciplinary symposium on 'Disorder and Order' in 1981, with an impressive list of speakers including Ilya Prigogine, Kenneth Arrow, Ian Watt, Henri Atlan, Isabelle Stengers, Cornelius Castoriadis, Michel Deguy, Heinz von Foerster and Francisco Varela. The symposium started a 'Program of interdisciplinary Research' at Stanford University, which would make multiple attempts to cross the 'Northwest Passage' (Serres 1980b). Together with Domenach, Dupuy also founded the 'Centre de Recherche et Epistémologie Appliquée' (CREA) in Paris, partly focused on Girard's work. Girard himself would be grateful to Dupuy, seeing him as 'the catalyst for many things' (Girard 2008, 42), especially making him 'aware of the relationship between "chaos theory" and the mimetic theory' (Girard 2008, 41).

Whereas Serres labelled Girard the 'Darwin of the human sciences', Dupuy goes one step further, seeing Girard as the Einstein of the human science (Dupuy 2013, 44). What he found in Girard's work was not a local hypothesis about religion and myth, but a framework that helped to solve the general problem of the generation of order out of disorder. 'Girard formally resolves the difficulty in the same manner, exactly, as theories of self-organization resolve a formally identical problem in biology: the passage from the simple to the complex, from the undifferentiated to the differentiated' (Dupuy 1988, 75). Girard's framework describing an initial chaos of mimetic desires and violence leading to a social order surrounding a scapegoat is no different from the new theories in cybernetics formulated by von Foerster, Prigogine and Atlan (Dupuy 2011, 207).

Dupuy illustrates this general logic with the example of a game in which one person has to reconstruct through yes-or-no questions the (historical) story the others have picked out for her. Through iterations, a focal point (the intended story) is reached. However, there is a variant of this game where 'the others did not choose a story, they simply agreed on a perfectly arbitrary response convention, such as: if the question asked ends with a vowel, the answer is yes, otherwise it is no' (Dupuy 1982a, 115). In this case, no pre-existing focal point existed, but a new order is created through a simple principle and a random starting point.

A similar role is played by the scapegoat. Dupuy himself makes the parallel with Robert K. Merton's concept of the self-fulfilling prophecy (Dupuy 1982b, 237). In his original essay Merton (1948) gives the example of a bank run: triggered by the rumour that the bank will go bankrupt everyone runs to the bank and it in fact goes bankrupt. But what caused this original false belief? The original act did not intend to cause a bank run, but was a local and contingent event: a misheard conversation or a misinterpreted message. Whereas this act on its own is contingent, it is retrospectively labelled as the original cause.

Serres would endorse these points by Dupuy, but adds a number of features. First of all, for Serres, objects too can play a role in the stabilization and creation of social order. In fact, mainly objects durably stabilize social relations. We already encountered this in Chapter 4, but it can be read in the light of Girard's theories as well:

> Nowhere do I see the sacred without a sacred object, a war or an army without weapons . . ., an exchange without values. The object here is a quasi-object insofar as it remains a quasi-us. It is more a contract than a thing . . . . The social bond would only be fuzzy and unstable if it were not objectified. (Serres 1982a, 88)

Serres sees this as one of the main differences with animal societies: 'For an unstable band of baboons, social changes are flaring up every minute. One could characterize their history as unbound, insanely so. The object, for us, makes our history slow' (Serres 1982a, 88).[2] To understand social order one thus has to look at how objects often play the role that Girard attributes to the scapegoat. In other words: *quasi-objects are structurally isomorphic to scapegoats*. Both create and stabilize social relations. It is here that Dumézil enters the scene again, for there are, according to Serres, three main types of quasi-objects structuring our social relations, related to Quirinus, Mars and Jupiter.

Science and violence

We can now start to explore the real novelties that Serres's work present. Similar to Serres, the Group of Ten crossed the Northwestern Passage. But Serres innovates this crossover in a number of ways. One is, as we saw, that he sees quasi-objects as being structurally isomorphic to scapegoats. But secondly, Serres also combines Dumézil's tripartite system with Girard's theory, linking it to Quirinus, Mars and Jupiter. Thirdly, Serres will also bring science itself into the picture, linking it mainly to Jupiter, which will therefore be our main focus. Finally, Serres also connects it with a diagnosis of critique, which we will explore in a separate subsection. But let us first start with Quirinus and Mars, before going to Jupiter.

Quirinus and the economy

Girard himself saw modernity mainly as a period where mimetic desire was no longer contained by religion, but instead became confronted with an endless supply of mimetic models. 'What is distinctive in modern times is that the array of models to choose from is much larger and there are no longer class differences in terms of desire – meaning that any external mediation in modern society has collapsed' (Girard 2008, 61). But we have to be weary not to misinterpret modern society as 'a society totally in control of its meanings, an immediate and transparent community to itself' (Dupuy 1982b, 258). Social order always requires a (pseudo-)transcendent source that limits its violence. If not religion, than through something else.

To understand these pseudo-transcendences, Dupuy invokes the work of Ivan Illich. In the 1970s, Dupuy was concerned with the problem of escalating healthcare costs (Dupuy and Karsenty 1977). In Illich's *Medical Nemesis* (1975) he found an ally and would join Illich's seminar at the Centro Intercultural de Documentación (CIDOC), where he met Heinz von Foerster (Dupuy 1982a, 11). The central notion in Illich's work is that of *counterproductivity*: 'Each major sector of the economy produces its own unique and paradoxical contradictions. Each necessarily brings about the opposite of that for which it was structured' (Illich 1981, 10). Once again, this concerns the question of order and disorder, but this time how order can transform into a disorder on a second level: 'Institutions thus appear, when certain thresholds are exceeded, as producing the opposite of what they are supposed to produce' (Dupuy and Robert 1976, 55). Illich's own work can be seen as the exploration of this logic through a number of case studies, such as education (Illich 1971), transportation (Illich 1974) and medicine (Illich 1975). In each case, initial autonomous and individual acts produce a transcendent and heterogeneous system, which risks betraying these original intentions: an educational system that makes us dumb, a transportation system that slows us down, or a medical system that makes us sick.

Underlying all these systems, for Dupuy, is the economy – and so we come to the first of the three types of quasi-objects: Quirinus or economic quasi-objects, such as money. Serres sees these economic quasi-objects as one way in which mimetic violence can be put to a hold. 'Money will channel violence, it carries it along and substitutes itself for it' (Serres 1982a, 90). In a similar vein, Dupuy describes the economy in Girardian terms: 'At least in liberal utopia, the market is a protective device against violence: the prices to pay oblige the parties to consider the damage they cause to others' (Dupuy 1982a, 77–8). In *L'enfer des choses* (1979) Dupuy, together with Dumouchel, reconceptualizes economic relations according to Girard's mimetic theory:

> Mimicry, that is, the fact that men imitate each other in their desire, the fact that we only desire objects which are designated to us by others, the fact that desire is always the imitation of another desire, desire for the same object. This changes everything, because economics has always thought of the subject-object relationship in general as a straight line that connects the subject to the desired object. (Dumouchel and Dupuy 1979, 11)

Though Girard is often optimistic about how the economy can put a hold to violence, Dupuy is more ambiguous. Inspired by Illich, he is aware how an economic logic can produce new forms of violence and disorder. There is a similar ambiguity at work in Serres's thought. For instance, he ends his text on Bachelard (Chapter 3) with an enigmatic call for a return to Quirinus (Serres 1970, 45). He seems to suggest that, rather than science in function of war (Mars), we need a productive, constructive science, linking it to the fertility and craftsmanship of Quirinus.

However, at other places, Serres is more pessimistic, since economic quasi-objects only pacify by, simultaneously producing violence: either in the form of the creation of a new proletariat (Serres 1994, 237) or by environmental pollution (see Chapter 8). 'Quirinus, god of production, or Hermes, who presides over exchanges, can sometimes keep back violence more effectively than Jupiter or Mars, but they do so using the same methods as Mars' (Serres 1990, 15). We thus find in Serres the same ambiguity towards Quirinus that he had towards science in general: it can be both a way out of the logic of violence and a reaffirmation of it.

Mars and war

Similar to Quirinus, Serres regards war (Mars) as a potential pacifier of violence as well. Rather than the absence of law and order, Serres understands war as a specific legal situation. War situates itself within specific parameters, agreed upon by both parties, about which actions are allowed, who can participate and when it ends. Thus, war is a way to restrict endless mimetic violence: 'When everyone fights against everyone, there is no state of war, but rather violence, a pure, unbridled crisis without any possible cessation, and the participating population risks extinction. In fact and by law, war itself protects us from the unending reproduction of violence' (Serres 1990, 13–14). Similar to communication (see Chapter 4), war can only exist by the exclusion of a third, limitless violence.

To clarify this logic, Serres centres his book *La Guerre mondiale* (2008a) around a story of a bar fight. First only two individuals fight and it is uncertain who started it, although soon others join. If left unchecked, the whole bar will soon be engulfed. But Serres imagines that one could, like a film, play the whole event backwards. We then see people leaving the fight, sitting down, restoring peace:

> The second part of the story, the one I projected in reverse, I call miraculous .... Yet we owe our survival only to this reversal. Question: how to achieve it? Who can lower the waters of the Flood? First answer: the port police, suddenly exited the vans alerted by the owner of the bistro who was trying to limit the damage; legitimate violence, therefore, before which the brawlers immediately stop exchanging their horrors. (Serres 2008a, 33)

Once again only violence ends violence, but best linked to an entity, such as the state, with a certain transcendence and authority. However, the state itself does not fight, rather its *representatives* (the police) do. War, similar to the scapegoat mechanism, is a question of finding the appropriate placeholders, lieu-tenants, who can fight in the name of all and within restricted rules. Armies and weapons, the second type of quasi-objects, are representatives, made to restrict violence:

> Weapons stop the battle or end it, they are only rarely used for fighting, spectacle or combat sport aside. Weapons are a freezing agent of violence, they are not necessarily its outbreak. . . . The stronger one presents arms and the weaker one runs away. Yes, arms are presented like monstrances. Hiroshima: the bomb ripped up the vanquished and, since then, has been getting displayed. (Serres 1982a, 88–9)

However, these clear rules of war seem to have disappeared. Mars has 'died', as Serres sometimes suggests (e.g. 2001, 293). But what then to make of all contemporary forms of violence, which have far from died out? Serres suggests distinguishing 'violence as *war*' from 'violence as *terrorism*'. In the latter case, violent acts are present without legal principles to contain them. It is unclear who the enemy is, what is allowed and when it will stop. 'In this case, no one knows who to negotiate with. Is there even some organized group that commands or sponsors these unpredictable attacks? No one can decide' (Serres 2008a, 96). Terrorism is a problem of representation, when the representatives become opaque or simply disappears: 'The lawlessness of terrorism begins as soon as I don't know who you are, where you are, what you are up to, where and when. The enemy is hiding under the possible friend' (Serres 2008a, 99).

Serres suggests that this shift from war to terrorism 'begins with France, with the Terror, with Napoleon; the fascination of Clausewitz bears witness to this' (Serres 2008a, 123). This seems to be a direct reply to Girard's *Achever Clausewitz* (2007), which reads von Clausewitz's *On War* through the lens of mimetic violence. Girard saw Clausewitz as a reflection of a time where the traditional rules of war were disappearing. 'Clausewitz was present at a decisive point when the shackles were broken. He saw violence rising under the increasingly meaningless surface of events' (Girard 2007, 91). Similar to Serres, Girard links this to the French Revolution, leading to further conflicts between France and Germany, ultimately leading to the two World Wars and all that followed. The result was an escalation of violence which dismantled the traditional legal framework of war, bringing us in an era of terrorism. In that sense, though violence is far from gone, this second type of quasi-objects seemed to have lost its power.

Jupiter and law

Serres saw war as a legal contract, restricting violence, which brings us to our third type of quasi-objects, embodied by Jupiter. The link between war and legality is not surprising, since Serres often stresses how modern jurisprudence can be traced back to its origins in legally framed conflicts, such as duels. According to Serres, 'the ordeal or the judicial duel bears, within justice-in-the-making, the trace of ritual sacrifice: a transitional stage, not unexpected, between the sacrificial rite and the law' (Serres 2008a, 85).

We can find this claim more articulated in the work of Serres's friend and colleague, Michel Foucault. In his 1973 Rio de Janeiro lectures Foucault aims to trace back the modern methods of inquiry and examination, linked respectively to the natural and the social sciences, to early evolutions in penal law. Foucault's third lecture maps the history of Germanic law, which functions, according to another model, remarkably close to the intermediary forms of jurisprudence that Serres described, the *testing game*. 'Germanic law did not assume an opposition between war and justice, or an identity between justice and peace; on the contrary, it assumed that law was a special, regulated way of conducting war between individuals and controlling acts of revenge' (Foucault 1973, 35).

Foucault describes a number of different forms this testing game could take. First of all, the social test, 'when a person was accused of murder, he could completely establish his innocence by gathering about him twelve witnesses who swore that he had not committed the murder' (Foucault 1973, 37). There was no need that these witnesses actually saw what happened, only their social standing mattered. Secondly, the verbal test, where the individual could win the game if he or she was capable to recite a number of formulas. Thirdly, 'the old magico-religious tests of the oath. The accused would be asked to take an oath and if he declined or hesitated he would lose the case' (Foucault 1973, 38). The gods would then proclaim the winner and strike down the loser. Finally, there were the ordeals: the bodies of the accused would put themselves at risk and the result was read as a sign of their guilt or innocence. To give just one example: 'The accused was required to walk on coals and two days later if he still had scars he would lose the case' (Foucault 1973, 38).

Girard similarly sees modern jurisprudence as a continuation of a sacrificial logic, fulfilling the same role of containing violence, but in a different way. 'Instead of following the example of religion and attempting to forestall acts of revenge, to mitigate or sabotage its effects or to redirect them to secondary objects, our judicial system rationalizes revenge and succeeds in limiting and isolating its effects in accordance with social demands' (Girard 1972, 22). Modern jurisprudence allows for direct punishment of violence, but limited to one trial only.

Dupuy has interpreted the use of voting ballots in the same light, which can be seen as a quasi-object around which society sustainably organizes political conflict. Violence is restricted to verbal debate and campaigning, with a clear-cut endpoint on election day: 'Like many religious rituals, the rite of voting involves two phases: a rivalry is first established, and then transcended in such a way as to bring forth an overarching consensus – the "will of the people" – that will guarantee social order' (Dupuy 2013, 104–5). This argument comes close to the agonistic view of democracy of Chantal Mouffe (2005), who similarly argues that democracy offers an acceptable way to pacify unavoidable political antagonisms. But Serres's notion of quasi-objects offers us a tool to also look at the role of non-humans, such as voting ballots, that are used to achieve this goal.

However, the most innovative element of Serres's view is his expansion of this argument to the realm of science, which he places under the banner of Jupiter as well. First of all, because science nowadays has taken on a number of properties, traditionally associated with religion:

> we give increasing importance to a third attribute of Jupiter, that of knowledge, of which Georges Dumézil speaks little. If history shows anything, Western history of science shows that science and law derived from religion. In traditional societies of this cultural area, the magi, druids, pastors, priests, clerics . . . in short, Jupiter, monopolized, in fact, knowledge and teaching for a long time. Nowadays it's the opposite, with the scholars forming a Church, with its dogmas, its dignitaries and its heretics, its hagiography and its rites. (Serres 1994, 222)

Secondly, one can read the history of science as a history of trials: of scientists being judged and of scientists judging society: 'the trial of Zeno of Elea, of Anaxagoras of Clazomenes, the condemnation and death of Socrates, the torture of Abelard, the burning of Giordano Bruno, the trial of Galileo, the beheading of Lavoisier, the tragic suicides of Boltzmann and Turing, etc.' (Serres 1994, 223).

Thirdly, science too follows a sacrificial logic of purification: expel one element in order to preserve, even constitute, a social order. 'Tell me what you're excluded, I'll tell you what you think. Things outside science, or the memory of the history of science, always teach us excellently about what passes for known' (Serres 1994, 93). This concerns not only the reduction and neglect of certain phenomena in the world but often also the exclusion of social groups or environmental concerns. In that sense, the informational cost (see Chapter 4) also gains a new meaning. Knowledge is never free, but requires a price of violence:

> Classical philosophy never calculated the cost of knowledge, thought or actions: it prejudged them free. It lived in the light world of grace and the given. However, as soon as work appears, everything goes through the martial law of price. Of Greek and Roman birth, philosophy presupposes that slaves pay with the sweat of their bodies the freedom of thought and action of those who live on leisure. (Serres 2001, 185)

For Serres, the exemplary embodiment of this logic remains the sacrifice of Hiroshima and Nagasaki to save the rest from war. Thus science does not end violence, but proliferates it in novel ways. Science is still reigned by death: thanatocracy. Elements of this scientific and technological pessimism are also found in Girard (1972, 240), but especially in Dupuy: 'It is my profound belief that humanity is on a suicidal course, headed straight for catastrophe' (Dupuy 2013, 21). At the same time, Dupuy explores nuclear weapons as pseudo-transcendent limits of violence: 'The very existence of nuclear weapons, it would appear, has prevented the world from disappearing in a nuclear holocaust' (Dupuy 2013, 16). Violence can be stopped only by a quasi-object, in this case the atomic bomb, as 'the absent – yet radiant – center from which all things emerge; or perhaps, to change the image, a black – and therefore invisible – hole whose existence may nonetheless be detected by the immense attraction that it exerts on all the objects around it' (Dupuy 2013, 175).

The problem, however, is that this only works insofar as the atomic bomb is perceived as a real and transcendent risk. Dupuy illustrates this through the following anecdote:

> In June 2000, meeting with Vladimir Putin in Moscow, Bill Clinton said something amazing. His words were echoed almost seven years later by George W. Bush's secretary of state, Condoleezza Rice, speaking once again to the Russians. The antiballistic shield that we are going to build in Europe, they said, is only meant to defend the United States against attacks from rogue states and terrorist groups. Therefore rest assured: even if we were to take the initiative and attack you first with a nuclear strike, you could easily get through the shield and annihilate us. (Dupuy 2013, 186)

Nuclear deterrence only works if it is perceived as a transcendent, almost inevitable thread. Dupuy therefore pleads for 'enlightened doomsaying' (Dupuy 2002), which 'invites us to make an imaginative leap, to place ourselves by an act of mental projection in the moment following a future catastrophe and then, looking back toward the present time, to see catastrophe as our fate – only a fate that we may yet choose to avoid' (Dupuy 2013, 33). Only by letting these catastrophic weapons function as transcendent quasi-objects do they contain violence, though simultaneously risk producing it.

This 'more archaic prehistory' of science is also found in other examples besides nuclear weapons. The most provocative one is found in *Statues* (1987), where Serres compares the deathly explosion on 28 January 1986 of the *Challenger* space shuttle at Cape Canaveral with an ancient sacrificial ritual in Cartago, centred around the incineration of humans in a gigantic brass statue of the god Baal. Both events involved 'statues' serving as quasi-objects, around which collectives organized themselves. There are more similarities, such as the role of specialists (priests or scientists), the huge societal cost, the large crowds and the repetitive nature, either by replaying the same images on television or by the recurrent nature of the ritual (for a detailed analysis, see Watkin 2020, 146–8).

Central for Serres, however, is the common element of the *denial* of violence. In the case of the Baal ritual, bystanders, who heard screams coming out of the fire, ascribed them to sacrificed animals. Similarly, the Challenger explosion was typically dismissed as an unintended accident. In contrast, Serres claims it was an essential element of science, since we can statistically predict that such accidents will occur, similar to how we know that driving cars will imply traffic accidents (Serres 1987, 9). We accept these sacrifices, and find them necessary in order for our society to function, similar to how mythical societies found scapegoats necessary to preserve social order. Endorsing our current way of doing science means accepting these sacrifices. Worse even, we deem them necessary, otherwise our society would fall apart. 'An ineradicable radical evil: the Carthaginians thought that they couldn't avert their sacrifices and the murder of their children in the same way that we demonstrate, by the calculation of probabilities, our inevitable errors or accidents' (Serres 1987, 13). Scientific quasi-objects such as the Challenger function as quasi-objects that stabilize our social order, while ambiguously also produce new violence.

As we will explore in the next chapters, since *Le contrat naturel* (1990) the ecological crisis has come to the foreground as well. Serres conceptualizes this as a literal world war, a war with the world itself (Serres 2008b, 109). We will return to the topic, but relevant for our story here is that Serres wonders whether mimetic violence could help us to understand and avert the ecological crisis. Can we conceptualize the Earth itself as a scapegoat, seen both as the source of violence – in the form of environmental catastrophes – and as sacred – the 'Grand Fetish', as Auguste Comte called it, or Gaia as Latour and Stengers now argue for? For Serres, the world war could be a way to channel mimetic violence among humans to a collective effort to struggle against environmental catastrophes: 'Peace with the world requires peace among humans. We will be saved from the apocalypse if and only if humans of all countries unite without borders to have the world as one partner' (Serres 2008b, 112).

In parallel with Girard's claims about Christianity, pre-Christian religions perhaps 'contained scraps of truth, no longer with regard to collective relations, but with regard to the world as it is and the global threats it shows' (Serres 2008a, 159–60). We laugh with how these ancient religions ascribed powers to natural forces and expressed reluctance to intervene in them – things that have become uncannily plausible in the Anthropocene. Serres therefore proposes to reconceptualize the world, following Auguste Comte, as 'the Great Fetish':

We then make the World, transhistorical victor-victim, both a weak, innocent and formidable divinity: the god of our ecologisms, the one I call, here, the Great Fetish.... Assessment: first a powerful enemy, accused, a target of indiscriminate violence, then almost put to death, finally deified, the objective world goes through the same stages as the subjective victim. The object turns into a subject. (Serres 2008a, 178)

Jupiter represents not only religion but law as well. Therefore Serres also proposes to integrate nature into the legal sphere, through his proposal of a natural contract. The current conflict with the world still remains in the non-legal sphere, that of terrorism: 'As a start of a possible law for the World, the Natural Contract aimed to transform the common piracy of exploitation into a conscious war, thereby being able to conceive a peace there. It did not constitute a final treaty, but an initial condition' (Serres 2008a, 169). This double dimension, of religion and ecology, will be the topic of the next two chapters.

Myth and critique

One additional mode in which science implies violence is that of critique. Modern critique often boils down to an accusation: you still believe in X. Critique thus claims the right to criticize everything and anyone without raising the question on what ground it has this right. 'It asks questions and remains suspicious of the answers, it never asks itself whether it has the right to act as it does' (Serres 1985, 43). According to Serres, the result is an unbeatable and boundless game of always unmasking one's opponent while simultaneously protecting one's own back against accusations by others:

> The hypocritical method consists in always placing oneself behind, and this immediately creates a queue. One must therefore get quickly behind the last person in the queue, stand behind the last one whose back can still be seen, then hide one's own back for fear of being caught in turn by someone who has understood the game. (Serres 1985, 43)

Inspired by Michel Foucault, Serres associates critique with surveillance and social science. We already encountered Foucault's genealogy of both the truth regimes of inquiry, in the natural sciences, and of examination, in the social sciences. Foucault traces the regime of *inquiry* back to the early history of penal law, when the aforementioned Germanic model was replaced. Germanic law disappeared mainly because of an accumulation of power in the monarch, who introduced himself as a third party in each dispute but 'the king or his representative, the prosecutor, could not risk their own lives or their own possessions every time a crime was committed' (Foucault 1973, 44).

Instead, it fell back on an already existing, but marginal alternative, found in the Carolingian Empire and the church: 'That method was called *visitatio*; it consisted in the visit the bishop was officially required to make in traveling through his diocese, and it was later adopted by the great monastic orders' (Foucault 1973, 46). It consisted of a general questioning of all that happened in the bishop's absence, and the possibility of an *inquisition specialis* if there had indeed been a transgression. While first being implemented in penal law, soon enough 'judicial inquiry spread into many other areas of social and economic practice and domains of knowledge' (Foucault 1973, 49).

But, according to Foucault, another model was developed in the nineteenth century, which became dominant in the social sciences, called *examination*. This part of Foucault's story is often summarized under the label of panopticism (Foucault 1975). Another form of truth or knowledge was produced, not based on a reconstruction of an original event, but

> a knowledge characterized by supervision and examination, organized around the norm, through the supervisory control of individuals throughout their existence. This examination was the basis of the power, the form of knowledge-power, that was to give rise not, as in the case of the inquiry, to the great sciences of observation, but to what we call the 'human sciences' – psychiatry, psychology, sociology. (Foucault 1973, 59)

Serres will copy this distinction, preferring inquiry over examination: 'Surveillance and observation. The human sciences keep watch, the exact sciences observe. The first are as old as myths; the others are born with us and are as new as history. Myth, theatre, representation and politics do not teach us to observe, they commit us to surveillance' (Serres 1985, 39). In contrast, inquiry is seen as free from this risk: 'Detach yourself from notions of winning or losing, be indifferent to victory or loss, you will enter into science, observation, discovery and thought' (Serres 1985, 44).

The critical attempt to purify science from all myth does itself not breaks with, but rather continues the archaic mythical gesture of finding a scapegoat to purify society. There is a mythical infinite regress at work in critique: 'philosophies which draw on the human sciences try to find sites which, in the final analysis, escape criticism, the last link of the chain, or the end of the queue' (Serres 1985, 44). Here Serres refers to the myth of Argus Panoptes (see Chapter 3), which embodies a game of deceit, where one always tries to trick the other: Zeus cheats on Hera and hides Io as a cow; Hera therefore tricks Zeus through Argus; Zeus, in his turn, tricks Argus through Hermes and so on. The same logic applies to critique: one always tries to outsmart the opponent. 'The human and social sciences describe theories even more underhanded than fraud, more duplicitous than cheating, in order to outsmart their object. Here everything becomes possible; a cow is a woman or a god a bull, even the identity principle is unstable' (Serres 1985, 43).

In opposition to this, Serres and Girard take an anti-critical stance, embodied by two characteristics. First of all, in an alliance with a certain realism:

> Without being able to prove it, I believe . . . that there exists a world independent of men. No-one knows how to demonstrate the truth of this proposition, which we might like to call realist, since it exceeds language and thus any utterance which might demonstrate its proof. Realism is worth betting on, whereas idealism calls for demonstration. (Serres 1985, 102)

Similarly, Girard states that he has 'always been a realist, without knowing it. I have always believed in the outside world and in the possibility of knowledge of it' (Girard 2008, 28). This realism is mainly a product of restraint: do not simply dismiss a phenomenon as unreal, a fetish, a political bias, a product of class interest. Do not reduce objects to the quasi-objects of Jupiter, Mars or Quirinus.

Secondly, it is linked with a certain naiveté. As Latour states, 'Michel Serres is naive and gullible beyond description' (Latour 1987b, 83). Girard as well endorses this label: 'philosophical naïveté is a definition that suits me' (Girard 2008, 33). This naïveté shows itself in the fact that both of them, at least initially, take mythical texts at face value, believing that they embody valuable anthropological lessons. 'Often despised by theorists, their imagination nevertheless seemed often to go deeper to me, towards anthropological truth, than many documents of history or philosophy' (Serres 2008a, 20). Similarly, for Girard 'religion is a true human science' (Girard 2008, 172).

Nonetheless, others have still problematized some aspects of Girard's philosophy, which still seems to fall under the banner of critique. First of all, Latour claims that Girard's own theory is still part of the sacrificial logic, 'since he accuses objects of not really counting. So long as we imagine objective stakes for our disputes, he claims, we are caught up in the illusion of mimetic desire. It is this desire, and this desire alone, that adorns objects with a value that is not their own' (Latour 1991, 45). In other words, Girard is unable to think about objects in any other way than projection screens of our mimetic desires.

Secondly, Jean-Pierre Dupuy and Henri Atlan have argued against Girard's theory's need for the element of misrecognition (*méconnaissance*): people cannot realize that they act according to a scapegoat mechanism. If they do, it would lose its effect. As an alternative, Dupuy relies on Atlan's epistemological framework of self-organization theory (Atlan 1988), which escapes this necessity of misrecognition by acknowledging the possibility of observation at multiple levels: 'observing at two different levels the social system of which they are a part, the actors can very well see, from the "exterior", the arbitrariness of the process of social differentiation, while still giving it meaning from the "interior"' (Dupuy 1988, 80). We can refer to a helpful example by Serres of a telephone call at a banquet:

> At the feast everyone is talking. At the door of the room there is a ringing noise, the telephone. Communication cuts conversation, the noise interrupting the messages. As soon as I start to talk with this new interlocutor, the sounds of the banquet become noise for the new 'us'. The system has shifted. If I approach the table, the noise slowly becomes conversation. (Serres 1980a, 66)

Whereas the conversation seems pure noise from an external point of view, once one shifts to an internal point in the system, the conversation regains its meaning. Dupuy himself gives the example of reciprocity in primitive exchange. Would a gift economy be destroyed once we tell the actors that they are in fact not really offering a gift, but always expecting something in return? This seems not to be the case. For instance, we are all aware of the fact that holiday gifts are not really gifts, since one has to repay in kind. We seem aware of this fact and install a number of procedures to avoid its official recognition, such as implicit rules on when and how to give something in return. Knowledge will not destroy the ritual. The ritual would only be destroyed if this implicit knowledge is explicitly institutionalized, making the internal perspective impossible: 'it is not knowledge which is incompatible with totalization, but its explicit expression and institutionalization' (Dupuy 1988, 91). The element of misrecognition, which still echoes critical unmasking, can thus be abandoned.

Conclusion

Though we started this chapter with Serres's analysis of nuclear weapons, we saw that this thanatocracy is more fundamental and called for an anthropology of science, which aimed to map this logic of violence. As Serres often repeats, 'science is followed, at a constant distance, by its own anthropology' (Serres 1987, 18). It was through Georges Dumézil and René Girard that Serres came to his anthropology of science, aimed to map this thanatocracy:

> Comparing the two comparative histories of religions therefore leads to reducing the [three] functions to one function or three gods to the only one and to showing the universality of sacrifice.... Abominable and present, this universal demands always and everywhere the death of men, in large numbers, in battles, having them, the production and circulation of goods. When my distant youth left epistemology, I called it *Thanatocracy*. (Serres 1994, 240)

Boundless mimetic violence is contained through three types of quasi-objects, embodied by Dumézil's tripartite system: law-religion-science (Jupiter), war (Mars) and economy (Quirinus). In each case, these quasi-objects paradoxically contain violence in two ways: they not only limit violence but also produce new forms of them. The atomic bomb is exemplary in this regard, since it both limits violence (nuclear deterrence) and magnifies it (nuclear destruction).

Is there an alternative? In his text on thanatocracy Serres suggests a strike by scientists, refusing to contribute to this mimetic spiral (Serres 1974a, 103). At other places, Serres suggests a new Hippocratic Oath for scientists: 'Today, therefore, we must rewrite an oath generalized to all sciences, since all scientists are faced with creative responsibilities' (Serres 1994, 245). More ambitiously, Serres dreams of an alternative, a science which is free from this tripartite system:

> We will seek another, different social universal. Right away we will have to discover a new object. The money is using itself up, the weapons are at their maximum, the fetishes are dead. I know what the object of science is. But we must find a different object, if we want to survive. Sacred objects stop violence, for a time only.... Armies stop violence as well, for a time only. The old terrifying god squatting behind the nuclear flash is rapidly wearing out. Money still stops violence, but also for a time only, for it runs away from it. Inflation has reached the quasi-objects. Can one imagine a different object of science, can one conceive an object of love? (Serres 1982a, 91)

What this alternative might be is not always so clear. But a recurrent theme in Serres's scattered remarks is the use of religious language. At several places, for example, Serres links it to the notion of rejoicing [*réjouissance*]: 'There is no knowledge without rejoicing' (Serres 1970, 86). In the final two chapters the aim is therefore to explore this alternative, through the themes of religion and ecology.

7

The secularization of science

Introduction

In this chapter I want to explore some themes on science and religion, which can help us to understand how Serres aims to escape or limit the violence produced by it (see Chapter 6). The aim is thus not to give a full overview of Serres's views on religion. Rather, what interests me is how for Serres a number of religious notions are seen as fertile to also understand science. We thus want to explore how Serres, and a number of authors who come close to his thought, offer us an interesting view on what science is and how we should relate to it.

As a starting point, let us note how Serres regularly applies secularization metaphors to science, arguing that our age 'is not yet secularized in relation to it' and still exercises on many scholars 'the pull of the sacred. The whole thrust of the epistemology or history of science can be read in this light' (Serres 1985, 334–5). We will see that Serres is not alone in mobilizing this religious metaphor of secularization to science, but that this has become a common trope in sociology of science. Does this then mean that Serres is merely a case of someone who wants to debunk our devotion to science?

The goal of this chapter is to argue that this is not the conclusion Serres wants to draw, even if he subscribes to a certain secularization metaphor applied to science. In the previous chapter, we saw how Serres is critical of a strong sociological perspective and its implied critical attitude. The secularization metaphor seems to imply such a critical stance: science is unmasked as simply a social phenomenon, a power struggle. There is nothing special or transcendent to science. This is not the conclusion Serres wants to draw.

That Serres's position is more complicated is shown by another metaphor that Serres uses to draw parallels between science and religion, namely that of a 'hot spot' [*point chaud*]: 'the places where, at such and such a moment, such and such another world manifests itself in this one; concrete images of virtual, intelligent, spiritual, inspiring, perhaps even dangerous contacts with this other reality' (Serres 2019, 14). One layer of these hot spot metaphors is what lies beneath: earthquakes, volcanic eruptions and so on. But Serres also associates hot spots with what is above, transcendent: places where meteors strike down, where the sun touches the earth, places of contact between the cosmos and our earth.

Serres develops this metaphor through a number of recurrent stories, such as that of the *gnomon* (see Serres 1989), the sundial as hot spot

> between the sun and the ground through light and shadow, as seen by our eyes, but above all between a vertical, material rod and a decodable knowledge, which I can call software [*logiciel*]; between the concrete on the one hand and the abstract on the other, the energy of light and the subtlety of information. (Serres 2019, 16–17)

A similar story is that of Thales, who in the shadow of the pyramids, finds a hot spot and founds mathematics (Serres 1993b). Thirdly, Serres often returns to the triple hot spot of Sicily: Archimedes, Empedocles and Ettore Majorana. Archimedes, maker of war machines, providing Syracuse with reflecting mirrors, burning away enemy ships by sunlight; Empedocles, throwing himself in the volcano, which refused to devour him, throwing back his shoe; and the mysterious disappearance of the mathematician Majorana, perhaps marked by how his work could contribute to the atomic bomb and thus our collective destruction:

> volcano, incendiary mirrors, a lucid and blind announcement of Hiroshima. . . . A reduced model of our world and its history, the Sicilian triangular manifests three hot spots. In short, will we turn our world into a hot spot today? For a long time we believed that the fires of science produced less violence than those of religion; you and I were wrong. (Serres 2019, 20)

Serres uses the 'hot spot' as a metaphor for scientific inventions, as the place where abstract theories and empirical reality meet: 'An invention therefore short-circuit a precise, virtual locality of mathematics with a defined phenomenon of the real world; a thread among the tissue which, virtually at least, unites equations and experiences. This short circuit still produces a hot spot' (Serres 2019, 22). The clearest example here, to which we will return, is that of Galileo:

> Galileo's most decisive and genuine invention lies in the connection he made between mathematics and experience. The Greeks had missed this point of intersection, so they could not develop an exact science of the world. Galileo, on the other hand, connects this equation with that manipulation. Thus, by a blinding and fruitful short-circuit between a virtual and formal world and the

real and perceived world, he announces modern science. His mathematical physics is breaking through a hot spot. (Serres 2019, 21)

To understand how science relates to society, Serres falls back on metaphors of transcendence: phenomena here are linked with what lies beneath or above. Now this opens a clear parallel to religious phenomena, of which Galileo is again a clear example:

> Now the Church of Rome taught the Incarnation of Jesus Christ, that is to say, the short-circuit, of a blinding light and bearer of Christian truth, between this real, incarnate world, on the one hand, and, on the other hand, a kingdom definitively separated from it. A hot spot, if there was one. By a similar gesture, did Galileo steal the dogma? (Serres 2019, 22)

Serres is not the first to make this connection between the transcendence of science and that of religion. In the 1960s a similar connection was made by Alexandre Kojève, who drew a parallel between the emergence of the hot spot of mathematical physics – where theory and phenomena met – and the Christian notion of Incarnation (Serres 2019, 22). In a contribution to a *festschrift* for Alexandre Koyré, Kojève argues that this Christian notion made modern science arise in the Christian West:

> If, as devout Christians assert, an earthly (human) body can be 'at the same time' the body of God and thus a divine body, and if, as the Greek scholars believed, divine (heavenly) bodies correctly reflect eternal relationships between mathematical entities, there is no longer any reason why we should not look for these [eternal] relationships here below as well as in heaven. (Kojève 1964, 303)

This metaphor thus seems to suggest that, for Serres, science and religion are closely intertwined, both linked to transcendence. Hence, a simple secularization thesis cannot be ascribed to Serres. But neither does Serres simply endorse a radical transcendent picture of science, since that would leave the transcendent violence we encounter in last chapter unquestioned. What I want to explore in this chapter is therefore how Serres position offers a fascinating intermediary position between these two extremes. In order to map the difference, it is meaningful to bring Serres into discussion with sociologists of science who also use this secularization metaphor.

Serres does not directly go into dialogue with sociologists of science and religion, but some of the authors influenced by his work have done so. A clear example of this is Bruno Latour. Although he already wondered in the past whether 'one could secularize Science without losing objective knowledge'

(Latour 2010a, 157), it is especially in his recent work on Gaia, that he claimed that 'the ecological mutation . . . obliges us to secularize – perhaps even to profane – all the (counter-)religions, including that of nature' (Latour 2015a, 179). This is especially clear in his third lecture, entitled 'Gaia, a (Finally Secular) Figure for Nature' where he describes Gaia as 'probably the least religious entity produced by Western science' and one that even 'may be called wholly secular' (Latour 2015a, 87). Stengers, in a similar vein, is critical of sociologists of science. Her aim is to develop a perspective that resists a purely sociological description of the scientific practice according to which 'we can henceforth enter his laboratory as if it were a windmill, open to all the influences of the epoch' (Stengers 1993, 42).

Let us therefore focus in this chapter on how these secularization metaphors are also used by Latour and Stengers, since they offer us a chance to explore the complexity of how Serres would link science and religion, and how a certain religious vocabulary and a specific take on transcendence can help us to understand religion as an antipode to the violence in science.

Secularization metaphors

In his book *Governance of Science* (2000), the sociologist Steve Fuller noted 'the profound historical irony that sociology has been both sanctifier and secularizer of science' (Fuller 2000, 99). The fact that sociology contributed to the sanctification of science is found in its origins: the work of Auguste Comte. For Comte the major problem was the societal gap left after the French Revolution: If religion was no longer there to hold society together, what would step in its place? Comte's positivism 'anointed the natural sciences the successors of the Roman Catholic Church as keepers of the key to the City of God on earth' (Fuller 2000, 99). For Comte, the secularization of society resulted in the sanctification of science.

However, recently a shift is taking place where 'we are in the midst of a second phase of secularization – that of science itself' (Fuller 2000, 100). In Fuller's view, sociology secularizes science through means of demystification. This process of 'secularizing science' entails depriving science from any transcendent or sacred position in society.[1] According to Fuller, this has been the main impetus in the 1970s of the sociology of scientific knowledge (SSK), which aimed to analyse scientific practices as social practices.

It is important to note that 'secularization' in this narrative is used as a metaphor, derived from the sphere of religion and applied to science. The metaphor of secularization thus aims to highlight to how SSK is eroding the 'sacred' and 'transcendent' status of science, replacing it by a 'secularized' view on science, with the connotation that it is more rational, grounded or empirical. The metaphor is thus mobilized, as we will see, as part of a rhetorical story in which traditional philosophy of science is dismissed as a naïve, pious yet blind admiration of science. SSK is portrayed as a disenchantment, freeing humanity from this idol of science. Of course, one might have reservations regarding the adequacy of the use of the metaphor and whether science and religion can be compared with one another in such a straightforward manner. Although these are legitimate concerns, I believe it is nonetheless productive to explore how sociologists have been using this metaphor, if we want to grasp Serres's specific take on it.

First, the goal is merely to understand how secularization metaphors are used in a variety of ways, without arguing for a single correct use. Secondly, it should be stressed that religion and secularization, although linked, are not the same. While science and religion are clearly different, for the authors discussed here 'secularization' refers to a logic behind certain societal shifts. In that sense it concerns the claim that a parallel logic is at work in science and religion, regardless of whether science and religion are similar, in essence or practice. In the third section of this chapter, therefore, the way this logic is understood in the study of religion will be explored by looking at established perspectives on secularization, especially within the work of Marcel Gauchet and how his work differs from sociologists of religion, such as Peter Berger and Thomas Luckmann. The central claim is that looking at how these philosophers of religion differ from their sociological counterparts can teach us something about how the philosophical perspective of Serres differs from that of sociologists of science.

In the fourth section, the manner in which sociologists of science have mobilized a similar logic to frame the history of science and their own position in that history is examined. Specifically, SSK will be examined, as well as a number of feminist authors, who have mobilized secularization metaphors in their own work. In the next part, we will then return to Serres and authors partly shaped by his perspective, such as Latour and Stengers. By putting these authors beside the earlier sociologists of science, it becomes clear that although they all mobilize secularization metaphors, they disagree on the nature and the implications of this logic of secularization. Here, precisely earlier differences between philosophers and sociologists of religion can shed a light on the work of Serres's distinction from sociologists of science.

The secularization of the West

The question of the nature and role of secularization in Western societies has been a prominent discussion within sociology of religion. Most famously, this has been conceptualized by Max Weber in his notion of the 'disenchantment of the world' (Weber 1991, 139). However, all too often this slogan is reduced to the claim that secularization implies a replacement of the falsehoods of religion by the truths of science (see Harrison 2017). It is indeed such 'subtraction stories', where religions are dismissed as false stories hiding an underlying reality that we now recognize because of scientific maturity, that are generally refuted in contemporary secularization narratives (Taylor 2007, 22). In this sense the claim that science causes secularization has been dismissed to 'the graveyard of failed theories' (Stark 1999, 269). Weber never intended such a claim. Rather, he defined disenchantment as the thesis 'that principally there are no mysterious incalculable forces that come into play, but rather that one can, in principle, master all things by calculation' (Weber 1991, 139). Or closely linked to this, that 'the ultimate and most sublime values have retreated from public life either into the transcendental realm of mystic life or into the brotherliness of direct and personal human relations' (Weber 1991, 155).

For Weber, this process is not caused by scientific insights debunking religious thought, but rather by a process of rationalization and intellectualization already at work within religion. Disenchantment is a process 'which has continued to exist in Occidental culture for millennia' and 'to which science belongs as a link and motive force' (Weber 1991, 139), but which is not exhausted by science. Such a process was already at work within theology, and even more generally in the codification of religion in rules and holy scriptures, exposing our worldviews 'to the imperative of consistency' (Weber 1991, 324). It is therefore also a practical process, where for instance inner-worldly asceticism is a crucial step towards rationalization, bureaucratization and modern capitalism (Weber 1985).

Sociologists of religion have picked up Weber's narrative, especially in the 1960s when a new generation of scholars, such as Peter Berger and Thomas Luckmann, took up Weber's work to criticize simplistic secularization narratives. Berger, following Weber, stressed how secularization did not start with science, or even with the Reformation, but 'begins in the Old Testament' (Berger 1967, 113). Berger therefore concludes that 'historically speaking, Christianity has been its own gravedigger' (Berger 1967, 127). Luckmann, on his turn, emphasized that secularization should be thought of as more than a mere emptying of churches, but is related to deeper societal transformations where 'autonomous institutional "ideologies" replaced, within their own domain, an overarching and transcendent universe of norms' (Luckmann 1967, 101). On its turn, such a focus on the underlying logic of secularization has been taken up by philosophers such as Marcel Gauchet in his *The Disenchantment of The World* (1985) and Charles Taylor in his *A Secular Age* (2007). Both authors are in fact in line with a broader sociological and historical consensus about secularization (Koenig 2016), but their work is nevertheless valuable for three reasons.

First of all, their work has proven to be crucial to engage philosophers into otherwise purely sociological debates. Secondly, these philosophers go beyond the mere negative task of criticizing and into the positive task of building an alternative narrative. This is also expressed by Taylor, in his preface to Gauchet's book, where he states that Gauchet 'argues, rightly I believe, that by never spelling out the big picture we have become unconscious of our ultimate assumptions and in the end confused about them' (Taylor 1999, ix). Thirdly, and most importantly, in contrast to sociologists, these philosophers of religion express a clear normative aspect, focusing on a range of tensions produced by secularization. It will be precisely this aspect that provides insight in Serres's specific take.

Although I will only outline Gauchet's point of view, it is perhaps helpful to introduce Taylor's distinction between three different ideas of secularity (Taylor 2007, 2–3). Secularity 1 refers to a decreasing relevance of religion in the public sphere, where for instance political discussions occur without reference to God or the holy scriptures. Secularity 2, on the other hand, entails a decline in religious belief and practice. Both these levels seem to be at work in SSK, as illustrated by the view of Fuller.

Despite the fact that both these processes are important, Taylor is especially interested in Secularity 3, referring to a shift in the background assumptions of a certain era, making it secular. It thus has to do with 'the conditions of belief. The shift to secularity in this sense consists, among other things, of a move from a society where belief in God is unchallenged and indeed, unproblematic, to one in which it is understood to be one option among others, and frequently not the easiest to embrace' (Taylor 2007, 3). According to Taylor, a focus on these deeper levels of secularization is also at work in Gauchet's perspective (Taylor 1999, ix–x). And as we will see, this dimension, linked with the normative aspect mentioned earlier, will be important to understand Serres. But let us first look to Gauchet into more detail.

Marcel Gauchet and the disenchantment of the world

In his own work Gauchet never speaks about secularization, but always about the 'disenchantment of the world'. The reason seems to be that Gauchet accepts secularization as a fact, but not as an explanation. Similar to Taylor, for Gauchet it is not a question of a decrease of belief or practices, but rather a shifting logic of the religious. What Gauchet is looking for is the underlying logic that leads to secularization, and he finds this in the idea of the disenchantment. But for Gauchet this notion is not exactly the same as for Weber. His definition is also different, namely as 'the impoverishment of the reign of the invisible' (Gauchet 1985, 3). Weber's notion of the disappearance of magical powers is thus an effect of this underlying shift.

Additionally, Gauchet radicalizes Weber's perspective in a number of ways. First of all, disenchantment spans the whole history of religion, going back to 'primitive' religions as well. For Gauchet the roots of disenchantment were already present within this original form of religion. Rather than looking at the history of religion as a progress and leading to the true religions of the book, Gauchet reverses the history. The purest religion is primitive religion, and the whole history consists in a process of departing from religion. In that sense, as Taylor also notes, 'Gauchet's story is not one of a development ... [but] a story of the breakdown of religion' (Taylor 1999, x). Later religions such as Christianity or Islam, 'far from being the quintessential embodiment of religion, are in fact just so many stages of its abatement and disintegration' (Gauchet 1985, 6).

Secondly, religion is also a more crucial element in our lives than most sociologists of religion claim. This has to do with two philosophical-anthropological claims made by Gauchet. First of all, Gauchet starts from the idea that what it is to be human, to be a subject, is defined by something outside of it, a certain alterity. What he means by this can be grasped in ideas such as the notion that language, social norms or even our biological bodies constitute us from the outside, forcing us to have a relation with elements over which we have no control. Gauchet calls this fundamental anthropological fact 'radical dispossession', and it is this fact that can give rise to religions, by linking this alterity with the invisible:

> Man is a being who, in any case, is directed at the invisible, and to whom demands are made by the other. These are orientations given in his original and irreducible experiences. . . . Man speaks, and he meets the invisible in his words. He experiences himself, irreducibly, as under the sign of the invisible. He cannot think that there is nothing else in him than what he sees, touches and smells. . . . Religions arise from these primal experiences. (Ferry and Gauchet 2004, 61–2)

The second, and related philosophical claim is that, because this alterity is linked to things outside of the subject, the structure of the subject also shifts in relation to the way this dispossession is shaped. Since religion has until recently been the dominant factor in the organization of society, the history of the subject and the history of religion are closely intertwined.

These starting points allow Gauchet to sketch his original historical narrative, namely one where the gap between the immanent/visible and the transcendent/invisible is progressively widened, until eventually the transcendence disappears out of view. I cannot and need not go into details of this narrative here, but some broad strokes can be highlighted. Where in primitive religions this gap was purely temporal, and the gods roamed over the same lands as we live in now, in later religions this transformed into a spatial gap, where God lives in a radically separated location. Gauchet links this to the rise of the state, which resulted in a separation of an elite, portrayed as spokespersons of the invisible, from the rest of society (Gauchet 1985, 14). Eventually in later religions, and especially in Christianity (and its Reformation), this led to a strong ontological divide between the visible and invisible.[2]

It is here that he focuses on elements already highlighted by other authors such as Weber, namely that within a world where the divine resides in a radically different and inaccessible place, humans are forced to focus on themselves and the earthly. But again Gauchet stresses the role of the subject, claiming that together with the constitution of an objective world, deprived of the divine, a subjective interior comes into being, creating the idea of a free-deciding individual. According to Gauchet there is a 'structural link ... between artificially appropriating the world and the political emancipation of individuals. Humans are initially free because they are alone before an empty and totally accessible nature' (Gauchet 1985, 70). The end result is the rise of new autonomous spheres of reason (philosophy and science), of political power (democracy) and of interactions with nature (economy and technology).

The consequence is, according to Gauchet, the 'end of religion'. This end is not a question of individuals losing faith or churches being empty. 'Leaving religion does not mean abandoning religious belief, but leaving a world where religion is a structuring element, dominating the political form of society and defining the structure of the social context' (Gauchet 1998, 11). Sociological criticisms that point at how in our contemporary societies new religions such as New Age are on the rise, therefore, miss their mark. As Gauchet puts it, 'this "return of the religious" seems to me to be anything but a return to religion' (Gauchet 1998, 29).

We have never been immanent

In the previous section we saw how Gauchet incorporates philosophical-anthropological theses in his work. This brings him beyond a mere descriptive history of secularization, as found for instance in the work of sociologists such as Berger or Luckmann, opening up a normative horizon. Gauchet mobilizes these theses to formulate a range of criticisms of our contemporary society. As stated before, these critiques will give us a clearer sight, in the next part of this paper, on Serres's point of view and how it differs from how the secularization metaphor is mobilized by other sociologists of science.

As we have seen, Gauchet starts from the fundamental assumption that subjectivity is defined by a form of alterity. Traditionally this element gave rise to religious practices, but 'the modern Western world's radical originality lies wholly in its reincorporation, into the very heart of human relationships and activities, of the sacral element, which previously shaped this world from outside' (Gauchet 1985, 3). In that sense, he claims that '[r]educing otherness does not mean eliminating the dimension of the other in the name of pure presence but transferring the other into immanence' (Gauchet 1985, 166).

Gauchet indeed ends his book with highlighting how this alterity is still at work in other 'secularized' forms, for instance in still popular distinctions such as those between 'appearance and truth, sensible and intelligible, immanence and transcendence, etc.'. (Gauchet 1985, 201). Gauchet gives a range of other examples, such as modern ideologies of communism, contemporary human right discourse, the focus on the unconsciousness, shifts in aesthetic experience, or the educational shift towards the 'open future'. But the most interesting one here is modern science,

> which clearly postulates the objectivity of phenomena, but simultaneously disqualifies any direct sensory observation of them, in favor of investigating the object's real properties, which it locates in the invisible. If on the one hand science expels the invisible from the visible (occult causal agencies), on the other it accommodates the invisible in the visible in a profoundly original manner, by installing an invisible certainty about its order at the very heart of the world, more certain than the world's appearances. (Gauchet 1985, 202)

Each of these practices tries to incorporate this alterity in an immanent secularized way, often without realizing it. Such a perspective allows Gauchet to diagnose a range of tensions within contemporary society, which fails to come to grips with this alterity. Part of Gauchet's project is to develop a new framework in which this alterity can be acknowledged, without going back to traditional religion. 'Let us say goodbye to the supernatural, but let us hold on to the reference to something outside of us that structures ordinary nature' (Ferry and Gauchet 2004, 114).

The secularization of science

We have seen how secularization is conceived by sociologists and philosophers of religion. This was done mainly in view of what will happen next: mobilizing some resources from these discussions about religion to shed a light on metaphorical claims about the secularization of science. Surprisingly this has not been systematically done before. This is especially remarkable if one thinks about how science often played a crucial role in discussions about secularization. This was already clear in the popularized version of Weber's secularization thesis, claiming that science pushed religion aside. Similarly, in Weber's work, both science and religion are shaped by the same process of rationalization. And it has indeed been noted that one could already conclude from Weber's analysis that science would be disenchanted in its own right.

But Weber does not discuss science as extensively as one would hope. Nevertheless, it is possible to make the argument that a similar logic at work in science has been described and endorsed by contemporary sociologists of science. Here we can follow Fuller's suggestions, namely that '[j]ust as sociology had contributed to the secularization of religion, science studies would contribute to the secularization of science' (Fuller 1999, 246). Fuller mainly refers to how from the 1970s onwards SSK came into being and aimed to analyse the content of science in sociological terms. According to SSK, the acceptance of Darwinism or Einstein's theory of relativity deserves a sociological explanation in the same sense that the acceptance of Social Darwinism or phrenology needs to be sociologically explained. True science is not something that transcends the social, but is rather part of it. In the first part of this section it will be argued that their work can be interpreted as a metaphorical secularization of science and discuss how they themselves have mobilized this narrative. But similar to the worries raised by Gauchet, the second part will show how Serres, and related authors such as Latour and Stengers, can be interpreted as a correction to a too strong reading of this secularization. They accept and endorse the secularization of science but argue that nevertheless some room for the alterity or transcendence at work in science must be provided.

Bringing science down to earth

In their own manifestos, introductions and histories several sociologists like to portray themselves as rebelling against a traditional position on Science. They argue against a view that allegedly sees Science as the ultimate source of all Truth, as a Method that reveals to us Reality itself. Of course, this attributed position might be more fiction than reality, but it nevertheless clearly serves a function in their self-understanding and a closer look at their claims and terminology is telling.

According to the sociologists, '[s]cientific knowledge does not carry a revelation of its own correctness along with itself' (Barnes and Edge 1982, 5–6). They thus dismiss the 'persistent idea that science is something special and distinct from other forms of cultural and social activity' (Woolgar 1988, 26). Or, differently put, 'scientific knowledge [is regarded] primarily as a human product, made with locally situated cultural and material resources, rather than as simply the revelation of a pre-given order of nature' (Golinski 1998, ix). As a result of their empirical studies 'the truth or falsity of scientific findings is rendered as an achievement of scientists rather than of Nature' (Pinch 1986, 20). Secularization metaphors thus serve a crucial role in their narratives.

This is particularly clear in the work of David Bloor, one of the founders of SSK. Rather than believing in the 'sanctity' and 'transcendence' of science, which makes a proper social analysis of knowledge 'beyond their grasp', Bloor looks at 'knowledge, including scientific knowledge, purely as a natural phenomenon' (Bloor 1976, 3, 5). Bloor even mirrors himself to Durkheim's sociology of religion, which had 'dropped a number of hints as to how his findings might relate to the study of scientific knowledge. The hints have fallen on deaf ears' (Bloor 1976, 4). Similarly, Barry Barnes and David Edge, two other founders of the field, dismiss all traditional analyses of science:

> Nearly all of these accounts of science are very heavily idealized, and represent the various utopias of our philosophers and epistemologists rather than what actually goes on in those places which we customarily call science laboratories in contrast, the present need is for a general description which treats the beliefs and practices of scientists in a completely down-to-earth, matter-of-fact way, simply as a set of visible phenomena. (Barnes and Edge 1982, 3)

This movement also inspired feminist scholars like Sandra Harding, who similarly opposes a traditional view of science which 'makes science sacred' (Harding 1986, 38). According to her, a down-to-earth sociological approach must be possible, allowing us to 'see the favored intellectual structures and practices of science as cultural artifacts rather than as sacred commandments handed down to humanity at the birth of modern science' (Harding 1986, 39). Donna Haraway echoes this critique and dismisses the 'conquering gaze from nowhere', associated with an allegedly disembodied, transcendent objectivity, which is in fact 'an illusion, a god trick' (Haraway 1988, 581–2). According to her, '[w]e have perversely worshipped science as a reified fetish' (Haraway 1991, 9).

However, none of the foregoing authors saw themselves as being antiscience. 'We see the sociology of scientific knowledge as part of the project of science itself, an attempt to understand science in the idiom of science' (Barnes, Bloor and Henry 1996, iix). Or similarly, Harding questions: 'Why is it taboo to suggest that natural science, too, is a social activity, a historical varying set of social practices? That a *thoroughgoing* and *scientific* appreciation of science requires descriptions and explanations of the regularities and underlying causal tendencies of science's own social practices and beliefs' (Harding 1986, 39)? The only element that they aimed to deny is the transcendent aspect of science, as if, in order to be true or valuable, science must be something radically beyond the social. This is what they mean by the sacred aspect of Science.

In the light of the history of secularization, such a strange tension between claiming to be scientific while being accused of debunking science is not surprising, on the contrary. Just as Gauchet shows how the departure from religion is prepared by religious arguments, movements and shifts (such as the Reformation), the secularization of science can be perceived as a step within science itself. Like the religious zealots who merely claimed to purify religion from its impurities, a similar logic is at work in science. As Joseph Rouse notes, the sociologists believe that 'philosophers' and scientists' faith in the distinctive rationality and progressiveness of the sciences is yet another irrational dogma that must finally succumb to (sociological) reason' (Rouse 1996, 7).

These sociologists of science thus suggest that one should reread the history of philosophy of science as one following a similar logic to the one exposed by Gauchet in the case of religion. Earlier positions which problematized 'naïve' forms of realism could be read as purification movements concerning science, claiming that Nature is too transcendent and far-away for our human concepts ever to grasp. At most we can approximate it, study its signs in the phenomena and experiences in our minds, but not Nature itself. Rather we end up in positions such as Karl Popper's fallibilism, where we can never grasp reality as such, but only make human attempts to approximate it. Popper indeed criticizes 'the doctrine that truth is manifest', referring to 'the optimistic view that truth, if put before us naked, is always recognizable as truth' (Popper 1963, 7). Such a direct contact with Nature is no longer possible. Truth or Nature is rather placed in a transcendent sphere, a 'Third World' (Popper 1968).

Just as in the case of religion, where a logic that progressively places God in a more and more transcendent spot makes Him irrelevant for our daily lives, so in philosophy of science putting Nature further and further away from our grasp raises the question of its relevance. Just as God became irrelevant for the profane life on earth, looking at science through the secularization metaphor leads to a view according to which Nature becomes irrelevant for the functioning of science, which instead received a secular explanation in social terms. The conclusion drawn by many sociologists of scientific knowledge, therefore, is that Nature has become irrelevant, since 'Nature can be patterned in different ways: it will tolerate many different orderings without protest' (Barnes and Edge 1982, 4). Similarly, critics such as Thomas Kuhn see this purification of the notion of Nature as a betrayal of the scientific understanding of science: 'It isn't that I think it's all wrong. . . . But you are not talking about anything worth calling science if you leave out the role of [Nature]' (Kuhn 2000, 317).

We thus see here a striking similarity to the logic at work in reformist movement in religion, which often see themselves as noble forms of purification, while simultaneously being demonized as the destruction of end of religion. In the case of science, this would mean the end of scientific rationality (to which they might contribute despite their best intentions). In that sense both the projects of the sociologists and feminists, as well of the critics who try to save parts of science from social influences, are part of the logic of a metaphorical secularization of science. But this logic has not been recognized as such, leading to fierce discussions in which questions such as 'Do you believe in reality?' become plausible to ask. But as Latour correctly notes, 'To ask such a question one has to become so *distant* from reality that the fear of *losing* it entirely becomes plausible – and this fear itself has an intellectual history that should at least be sketched' (Latour 1999b, 3–4). And this is precisely the so far unwritten history of the secularization of science.

A laboratory, not a windmill

Let us now turn to the extra element I stressed in the story of Gauchet, namely, how in his work there is also a critical perspective, opening up a normative correction of the secularization process. In the last part of this chapter I will argue that Serres, and related authors such as Latour and Stengers, aims to do a similar thing in the case of science.

In the introduction, I already quoted Stengers' worry that the secularizing gesture of the sociologists, results in a picture of a 'laboratory as if it were a windmill, open to all the influences of the epoch' (Stengers 1993, 42). Stengers is not opposed to the sociologist's perspective but stands rather sympathetic towards it. At the same time, however, she worries that sociologists are nevertheless overplaying their cards by indiscriminately denying any difference between scientific and other practices. Although the sociologists are correct concerning the fact there is no reified transcendent Nature, Stengers warns against confounding this claim with the claim that there is no transcendent element whatsoever in science and that everything is in the hands of the human scientists.

According to Stengers, there is in fact a form of transcendence or alterity at work in science, to which she refers to as an 'event'. Similar to Serres, and in order to illustrate this, she returns to the example of Galileo. Stengers is aware that much has already been written on the Galileo affair (see Scotti 2017). Her ambition is not to give a historical correct account of the Galileo affair, but merely to link an alternative narrative with the mythology surrounding Galileo, putting other aspects of the scientific practice into the spotlight. The aim of Stengers (and Serres as we will see) is essentially to use Galileo as an exemplar in order to tell a bigger story about what science in general is about.

In the narrative she tells about Galileo, she in fact comes close to the narrative about secularization. Similar to Gauchet, she focuses on the role of the nominalist logic in the dispute between Galileo and the church, which stresses the gap between our logic and the mind of God, which she sees at work in the argument of Cardinal Maffeo Barberini (the later Pope Urban VIII):

> If God had so willed, what seems normal to us would not be so to him, what seems inconceivable or miraculous to us would be the norm. . . . If no other difference between the imaginative and fictive world and our world can be legitimately invoked except God's will alone, . . . then any mode of understanding that is not itself reducible to the pure observation of the facts, and to the logical reasoning derived from the observed facts (bringing into play the principle of noncontradiction that even God respects), is of the order of a fiction. (Stengers 1993, 78)

According to Stengers, Galileo subscribes to this nominalist scepticism but exploits this to his own advantage. He accepts that any logic he would come up with to describe nature is a 'fiction', of which there is no guarantee that the world is actually so. But, so Galileo adds, he can make a difference, namely by citing the world itself, by mobilizing phenomena and let them authorize him to speak in their name. Galileo does something with the falling bodies which allow them to point at Galileo as their spokesperson. In that sense, Galileo's legacy entails 'the invention of the power to confer on things the power of conferring on the experimenter the power to speak in their name' (Stengers 1997, 165). Or to put it differently:

> The singularity of scientific arguments is that they involve *third parties*. ... What is essential is that it is *with respect to them* that scientists have discussions. ... [T]he scientists themselves only have influence if they act as representatives for the third party. With the notion of third party, it is obviously the 'phenomenon studied' that makes an appearance, but in the guise of a *problem*. For scientists, it is actually a matter of constituting phenomena as *actors* in the discussion, that is, not only of letting them speak, but of letting them speak in a way that all other scientists recognize as reliable. (Stengers 1993, 85)

And, although this might look like Stengers endorses the traditional transcendental role of Nature, it is in fact subtler. This is because Stengers adds that, although Galileo is indeed invoking a transcendent element here in the discussions (the falling bodies are intervening in the discussion), it does not follow that Galileo can identify this transcendent element in an unproblematic way (as Nature, for instance). In that sense Stengers wants to separate two elements, allowing one 'no longer to deny the differences scientists claim for themselves, but to avoid any way of describing them which implies that scientists have a privileged knowledge of what this difference that singularizes them *signifies*' (Stengers 1993, 67). Transcendence is thus present, but not as an authority to be unambiguously invoked, but as a problem.

Let us now return to Serres, since a similar argument is at work in *Le contrat naturel* (1990). In his own description of the Galileo affair, again making a point about science in general, Serres is playing with the different forms of transcendence being invoked. According to Serres, Galileo's famous last words 'and yet it moves!' must be interpreted as invoking a transcendental court, outside of society, namely the natural world itself:

> The cardinals decide and pass judgment in the name of canon law, of Roman law, and of Aristotle, the physicist jurist. To respond to them, Galileo tries to escape from these texts and conventions by positioning himself outside their laws: 'my kingdom is not of this world', he says, in substance, or, changing point of

reference: 'the world is not within the jurisdiction of this court'. He is appealing to a nonexistent authority. (Serres 1990, 84–5)

Galileo is thus constructing a new transcendent authority, Nature, which has, since then, become a dominant one. But what is at stake in Serres's book is a contestation of this authority, mainly for political reasons: the traditional transcendent Nature, so productive for the sciences, has resulted in a neglect for the violence as produced by the sciences, for instance in the shape of the ecological crisis (see the previous chapter). Or as Serres states:

> Science won all the rights three centuries ago now, by appealing to the Earth, which responded by moving. So the prophet became king. In our turn, we are appealing to an absent authority, when we cry, like Galileo, but before the court of his successors, former prophets turned kings: 'the Earth is moved'. The immemorial, fixed Earth, which provided the conditions and foundations of our lives, is moving, the fundamental Earth is trembling. (Serres 1990, 87)

The conclusion of Serres is not that we should get rid of all forms of transcendence, but only that the traditional narrative of this transcendence element as Nature should be revised. The ecological crisis shows that we are still confronted with a transcendent force, beyond our control, but one for which the traditional terminology is inadequate and which should therefore be rearticulated. In a sense, we find here is an echo of Taylor's critique of 'subtraction stories': sociologists run the risk of telling a subtraction story of dismissing previous illusions about science, replacing it by a purely immanent description of how science 'really' is. But a different model is also possible, namely where it is not a question of subtraction, but rather of articulating a different relation with the sciences and transcendence.

From the Great Fetish to Gaia

One way in which Serres aims to grasp that other model is to reintroduce talk about fetishes. Originating from the medieval Portuguese word '*feitiço*' (from Latin *facticius,* manufactured or man-made), it referred to a set of magical practices that the Portuguese encountered in West Africa. In 1757, the French lawyer Charles de Brosses coined the term 'fetishism' to describe how fetishes were not something particular to West Africa, but believed to be a primitive stage of all religions. A few years later, in 1760, de Brosses would publish his book on the topic, *Du culte des dieux fetishes.*

The way how fetishism is understood in traditional sociological analyses (ranging from Durkheim to Marx) is mainly as a subtraction story: a certain causality or set of powers is attributed to a mundane object, believed to be divine or magical. The task of the sociologist is then to debunk this: what seemed important is actually a mere fetish, something on which we falsely project the powers that are actually to be attributed to society and human labour. This use is, for instance, still at work in the previous citation of Haraway.

The way how Serres uses the notion deviates from this. For Serres, a fetish is indeed 'an object, in ivory or in wood made by humans, of which we know that it is humans who sculpted it and which is nevertheless the object of a cult' (Serres 2014, 260). Yet, according to Serres, whereas fetishes might have looked like foolish things in the past, this has changed in our contemporary society. Many of our contemporary crises remarkably share this confusion with fetishes: we created them, yet we do not fully control them, and even depend on them. We thus depend on the things that depend on us: climate change, economical conjunctures, global pandemics and the like. A form of transcendence is thus reintroduced. Or in Serres words: '*it no longer depends on us that everything depends on us*. This is the new principle or foundation of the new wisdom' (Emphasis in original) (Serres and Latour 1992, 172).

It is in this sense that the traditional vocabulary of subject versus object fails, according to which the notion of a 'fetish' referred to the traditional accusation of confusing a subject with an object, something active and powerful with something passive and powerless. Instead we must recognize that we are faced with quasi-objects: things that depend on us and on which we depend simultaneously. In other words: fetishes. Or, as Serres argues,

> today, the efficiency of knowledge have lost this somewhat obsolete idea that there is a subject, who is the master, and a passive object. The object in front of me is something other than a passive object, and I give it the status of a subject. In other words, it's a fetish! Here we are. This is the natural contract. So, it is really a scientific and philosophical question even if, traditionally, it comes to us from religions like those of the Aztecs or the Assyrians. (Serres 2014, 268)

In a very similar vein, Latour (1999b) and Stengers (2010) take up this issue as well. They, however, trade the term 'fetish' for that of *factish* (a contraction between fetish and fact). Latour's *Petite réflexion sur le culte moderne des dieux faitiches* (1996) clearly alludes to the original essay on fetishism by Charles de Brosses. But in contrast to the traditional scheme where fact is opposed to a mere belief, the factish suggests a different move:

> it is *because* it is constructed that it is so very real, so autonomous, so independent of our own hands. As we have seen over and over, attachments do not decrease autonomy, but foster it. Until we understand that the terms 'construction' and 'autonomous reality' are *synonyms*, we will misconstrue the factish as yet another form of social constructivism rather than seeing it as the modification of the entire theory of *what it means to construct*. (Emphasis in original) (Latour 1999b, 275)

Serres sticks to the term 'fetish', because he was inspired by Auguste Comte. As we saw in Chapter 1, Serres's relation to Comte was ambiguous. Though Comte was mistaken on nearly all scientific issues, Serres is still fascinated by Comte, mainly by his later work on the Religion of Humanity: 'positivism turns out to be weak where it is believed to be strong, precisely with regard to science and history, and strong in the areas where everyone condemns it, like an understanding of religion' (Serres 1989, 448). Though often dismissed as strange and ridiculous, Serres pleads to take the later Comte seriously. 'Because this part of his work is much more interesting than we think' (Serres 2014, 258).

Comte already used the notion of fetishism in his early work, from 1825 on. It was seen as an early stage of the theological stage, in his infamous law of the three stages. Though thus a stage to be overcome, it is nonetheless important to keep in mind that, for the early Comte, fetishism was a necessary step to get to positivism. 'Comte assigned fetishism a positive role as the first motor of progress' (Pickering 1998, 54). In his later work, however, fetishism started to play an even more important role, especially in his Religion of Humanity. For Comte, fetishism was a way to rethink the role of emotions and relations in our society, something a strict rationalist approach failed to acknowledge. Religion, for Comte and for Serres, thus raises important questions about how we relate to the world. It is in this context that Serres also picks up Comte's writings on the Earth as the 'Great Fetish':

> Comte forges a new 'positive Trinity', with three characters: the Great Being, which rules the Universe, the Great Fetish, which is Earth, and the Great Milieu, which is space. It is the second term that interests me and that I had found independently of it. To think of Earth as a Great Fetish is a brilliant invention. (Serres 2014, 259–60)

This fundamental insight is the one we encountered earlier, namely that we can and must understand our relation to nature and the Earth in terms of a fetish, the Great Fetish:

we depend henceforth of this enormous thing which now depends on us. This is how the Earth has become our Great Fetish. There are sometimes notions which, at the moment, seem absurd or crazy, but which, crossing time, reveal their deep truth a posteriori. The Earth has become Auguste Comte's Great Fetish: we make it ourselves, we shape it; and we depend on what we have produced. The more we produce it, the less we are in control of it! (Serres 2014, 260–1)

Within this context the secularization metaphor also pops up in Serres work, namely as the disappearance of the fetishistic relation to Earth, to the land. Religions such as Christianity were secularizers in the sense that they detach us from the land and the earthly: 'At one point, preachers came to tell the peasants that the Holy Land was not under their feet but in Jerusalem. Gigantic revolution! For the first time, the earth in general was no longer holy. The city of God was elsewhere. It was the beginning of the secularization of the earth' (Serres 2014, 264).

This brings us close to Latour, who similarly tries to capture our new ecological condition through a mythical figure, in his case the figure of Gaia. Serres himself will not use the notion of Gaia, but – besides that of the Grand Fetish – that of biogea (2010). We will return to both figures in the next chapter. Relevant here is how Latour describes Gaia explicitly through the metaphor of secularization, for instance when Gaia is described as 'a (finally secular) figure for nature'. At first sight this claim that Gaia is 'totally secular' (Latour 2015a, 106) seems to contradict the positions of Stengers and Serres and is rather in line with the sociological perspective. This is true insofar as Latour mobilizes the metaphor of secularization as a rhetoric device to expose concepts such as Nature or Science as ersatz religions. According to Latour, the idea that Science reveals Nature, which is the ultimate arbiter in all our disputes, is as religious an idea as those found in the traditional religions. The result of this is a 'gap that separates Science from the sciences, matter from materiality' (Latour 2015a, 211).

'Matter' is also a religious idea for Latour, where the plurality and agency of the world is reduced to an anonymous and sterile form, which paradoxically is claimed to be immanent but is transcendent and beyond our grasp at the same time. In that sense, Latour states (following Eric Voegelin), that modernity has not been secularized, but 'immanentized' and the moderns 'have no sort of possible contact with the terrestrial, since they can see in it only the transcendent, which would be trying awkwardly to fold itself into the immanent. And necessarily failing' (Latour 2015a, 204). To correct this we are in need of 'a conception of materiality that is finally worldly, secular – yes, non-religious, or, better still, earthbound' (Latour 2015a, 72).

But at the same time, for Latour, the metaphor of secularization is also suspicious, precisely because this notion contributed to the earlier confusion surrounding immanentization. 'To move forward, we would have to be able to establish a new contrast between, on the one hand, the terms *religious* and *secular* and, on the other, the term *terrestrial*. The terrestrial is *immanence freed of immanentization*.... The terrestrial is neither profane nor archaic nor pagan nor material nor secular' (Latour 2015a, 211–12). Or put differently, 'the "secular" is like non-alcoholic beer, it is the religious without religion. But Gaia goes further' (Latour 2015a, 87n38).

Latour tries to go beyond this by reconceptualizing what the religious is all about and by contrasting it with religion as traditionally understood. In this he takes inspiration from Serres, who defines religion not in opposition to atheism, but rather to negligence. 'Whoever has no religion should not be called an atheist or unbeliever, but negligent' (Serres 1990, 48). To be religious thus means to care about certain elements, a definition which for instance allows one to consider Galileo as religious, since he cares about his phenomena, and would resist against those who – by negligence – would reduce his laboratory to a windmill. So for Latour, '*there is no such thing as an irreligious collective*. But there are collectives that *neglect* many elements that *other collectives* consider extremely important and that they need to care for constantly' (Emphasis in original) (Latour 2015a, 152).

This religious element can be contrasted with both traditional religions and their secular counterparts. In this, Latour follows the work of Jan Assmann, who makes a (controversial) distinction between the original religions and what he calls 'contra-religions.' According to Assmann, the original religions worked through 'translation tables', allowing different gods to be defined through their actions and thus recognized by other collectives and cultures. Jupiter and Zeus could be placed on the same level, for instance. In this sense they align to the religious, as defined by Serres: they do not neglect the relations that matter for the other. The invention of Moses, however, was precisely to transform what was considered to be a vice, namely negligence, to become a virtue: one should be negligent towards the gods of other collectives. For there is only one true God. Assmann defines this as a contra-religion (see Assmann 1998).

What Latour adds to Assmann is the claim that not only must religions such as Judaism, Christianity, or Islam be seen as contra-religions but also our contemporary trust in Science and Nature are secularized contra-religions. 'From the true God fulminating against all idols, we have moved to the true Nature fulminating against all the false gods' (Latour 2015a, 157). In that sense, the metaphor of secularization breaks down, since it is part of the problem.

Precisely a genuine 'secularization' of our current predicament would require us to abandon all ultimate transcendent arbiters, both from religion and secularity.

'This is why it is so important, in my view, to try to face up to Gaia, which is no more a religious figure than a secular one' (Latour 2015a, 219). Gaia, for Latour, entails a 'terrestrialization', forcing us to recognize the elements that we do not want to give up without making them absolute in the form of an ultimate transcendence, including in an immanentized form, such as Nature. This would allow us to re-establish contact with the world, with materiality. It would allow us to acknowledge the things that move us, make us do more, as for instance was the case in Galileo and his phenomena, without reintroducing a traditional transcendent instance.

And although Latour does not refer to these discussions at all, it seems fruitful to interpret this in the lines of Gauchet. In the case of both religion and science, they stress how, although specific forms of transcendence are being problematized or have disappeared, other forms of transcendence still play a factor that should be taken into account. This seems to be a central disagreement that these disputes about secularization metaphors bring into the open. The tension between Serres, Latour and Stengers and the earlier described sociological programme centres on this: the metaphorical secularization of science as conceived by sociologists is untenable. Rather it is either a problematic purely 'immanent frame', without room for any transcendence or, even worse, the reintroduction a new immanentized figure, namely society. Such a perspective would be one where 'objects count for nothing; they are just there to be used as the white screen on to which society projects its cinema' (Latour 1991, 53).

It is possible to link these remarks on secularization to the earlier work of Latour. In *Nous n'avons jamais été modernes* (1991), Latour already pleads for a transcendence, but without a contrary. Instead, 'we get a single proliferation of transcendences. A polemical term invented to counter the supposed invasion of immanence, the word has to change meaning if there is no longer an opposite term' (Latour 1991, 129). Despite the different terminology, it is not in contradiction with the foregoing plea for immanence freed of immanentization.

Again following Serres, Latour calls 'this transcendence that lacks a contrary' *delegation*, referring to the idea that there is no ultimate essence or ground to fall back upon (such as the immanentized form of Nature), but rather what he elsewhere calls a 'mini-transcendence' in the form of 'a process, on a movement, a passage – literally a pass, in the sense of this term as used in ball games' (Latour 1991, 129). What things are is not determined by itself, but is rather always defined by the entities to which it is related. Galileo is shaped by his phenomena; the religious person shaped by those things he or she cares about. This is a form of transcendence, but not in the traditional radical kind. 'There is mini-transcendence; there is no maxi-transcendence' (Latour 2013a, 402). In that sense a true immanence would precisely have room for these mini-transcendences.

Conclusion

Serres is a religious thinker, but he mobilizes religious ideas and metaphors not just for their own sake but also because they can shed a light on the role of science in our society. In this chapter we tried to explore how Serres precisely does this. The starting point was a certain tension between, on the one hand, a reliance on a form of transcendence in science (in the form of hot spots) and, on the other hand, a set of secularization metaphors. The aim of the chapter was to highlight how Serres (and Latour and Stengers) developed a particular take on the question of science, secularization and transcendence.

To do so, a detour was made through the secularization narratives of sociologists and philosophers of religion. Crucially, in the work of Gauchet a particular normative aspect was present, allowing us to see how something similar is at stake in the debates concerning the secularization of science. Both Gauchet, on the one hand, and Serres, Latour and Stengers, on the other hand, want to go beyond their sociological counterparts, and both do so by mobilizing certain philosophical notions. Their goal is the same: to free up space for forms of alterity or even transcendence in a world that seems 'secularized'. In the case of Serres this was linked with his specific take on the Galileo affair and his interpretation of fetishism. The persuasiveness of such attempts, in both cases, depends on whether the introduction of certain philosophical notions into the debate is deemed acceptable.

At the same time, however, such similarities have their limits. A range of tensions between traditional secularization narratives and discussions concerning science and technology do exist. Since the goal here has been to stress similarities, disagreements and tensions were downplayed in the foregoing picture. Similarly, the question to what extent one can compare religion and science as easily as these authors would want to has been left aside. But to counter and limit this to a certain extent, let me end with highlighting a number of genuine tensions between Gauchet on the one hand and Serres, Latour, and Stengers on the other hand.

Let me start with what a Serresian perspective can say about secularization narratives. The main tension arises in relation to either the notion of the subject or the notion of science. This is especially clear in the case of Gauchet, who seems to state that what is ultimately at stake is purely a history of the subject. It is not a matter of how the world is transformed, and presents itself in different ways to us, but rather 'that of the internal constraints forcing us to present the question in this way' (Gauchet 1985, 202). One could thus criticize Gauchet for a tendency to reduce all that is at stake to 'the merciless contradictory desire inherent in the very reality of being a subject' (Gauchet 1985, 207). There is no genuine room for how we are transformed by things outside the subject. A certain typical modern focus on the subject is thus left unquestioned (see the next chapter).

A second issue is the notion of disenchantment. In Gauchet's story, modern disenchantment has problematized these transcendent shaping factors. From the perspective of Serres, one could point at an ambiguity at work here, namely between the claim that these transcendent factors have disappeared and the claim that we are only incapable of articulating them in our contemporary discourse. In the second case they would still be there, but we have become unable to give them due attention. Gauchet seems to endorse the first. But the reality of disenchantment might depend on how one looks at science and nature. Latour, for instance, denies 'that science has 'disenchanted' the world' and rather affirms 'that science has always *sung a quite different song* and has always *lived fully enmeshed in the world*' (Emphasis in original) (Latour 2015a, 72). For him, the problem is not science disenchanting the world, but rather a certain understanding of science that makes us insensitive to forms of enchantment that are still there, even in science. This is another unresolved tension.

At the same time, one could turn the tables, and criticize Latour. Gauchet, for example, questions a certain shift in 'representation' in contemporary democracy, and one could accuse Latour's own project, especially related to his Parliament of Things (see the next chapter), of unreflexively endorsing this shift. According to Gauchet,

> 'To represent' meant: transcending the differences between individuals or groups in order to show the truth of the community in the unity of her will. Now it means: showing the differences, guaranteeing their visibility in public space, ensuring that they are legible at all the moments of the political process, that they are not lost in the formation of public decisions. (Gauchet 1998, 119–20)

The problem is thus a disabling of democracy, since 'representation as a public staging of social diversity seems to become an end in itself' (Gauchet 1998, 121).

In the work of Latour and others, there is indeed a problematic tendency to celebrate pure diversity or plurality (see the next chapter). This precisely runs the risk that 'the endeavor to be readable in all its parts leads to a curious collective illegibility. . . . We finally arrive at a new contradiction, namely of a society that knows itself to the smallest details, but no longer understands itself as a whole' (Gauchet 1998, 126–7). We will turn to these issues in the next chapter.

8

The parliament of things and the Anthropocene
How to listen to quasi-objects

Introduction

In the last decade, it has become popular to speak of the dawn of a new epoch, the Anthropocene. Introduced in geology in the early 2000s, this new era refers to the moment when human activity started to have a significant or even a dominant influence on the planet (Crutzen 2002). There is still a discussion going on concerning the official recognition of this new label and the precise moment when the Anthropocene has started, ranging from the birth of agriculture, the Industrial Revolution to the first atomic bombs. More recently philosophers have started to mobilize this concept in their reflections on nature and technology as well (e.g. Morton 2014; Stiegler 2015).[1] Two clear examples of philosophers who have taken up this notion of the Anthropocene are Bruno Latour (2013a, 2014, 2015a) and Isabelle Stengers (2011a, 2015a). Michel Serres himself uses the notion of Anthropocene sporadically (e.g. Serres 2016, 43) but in general seems to avoid it. This might have to do with the fact that the Anthropocene still has an anthropocentric echo to it. As Watkin states, 'it registers only one direction of influence: human beings are changing the earth and its climate. The idea of fetish, by contrast, acknowledges the mutual influence of world and humanity on each other' (Watkin 2020, 367), But also in the case of Latour and Stengers, this notion appears as an extension of an approach they had been working on for decades, one deeply influenced by Serres.

Around thirty years ago, Latour was already calling for a new form of democracy, namely 'a democracy extended to things themselves' (Latour 1991, 142). In an age of climate change, nuclear disasters, GMO's, aids and economic crises, we cannot limit politics to subjects alone. These problems are neither pure politics, because they involve natural phenomena, nor pure nature, for they only exist due to the mediations of humans. We are thus in need of an 'object-oriented democracy' or a *Dingpolitik* (Latour 2005a, 14), which implies a rethinking of the role of science and technology and linking them with their political aspects. For Latour, this requires the creation of a 'parliament of things': a place where both humans and non-humans can be represented adequately (Latour 1991, 144). Here, Latour is inspired by Serres, who similarly called for a 'natural contract' (Serres 1990) and who later proposed a similar global institution, that Serres named WAFEL (referring to water, air, fire, earth and life):

> Let's conceive a new institution, which we could name WAFEL (Water, Air, Fire, Earth, Life), in which Homo politicus would welcome the elements and living things, non-appropriable quasi-subjects because they form the common habitat of humanity. At the imminent risk of death, we have to bring about peace between ourselves to safeguard the world and peace with the world in order to save ourselves. (Serres 2004, 188)

However, it is hard to grasp what Serres and Latour have in mind when introducing these institutions. At first sight, it seems to be problematic notions: incorporating things contradicts the history of philosophy. While philosophers used to see direct knowledge of the objective world as unproblematic, authors since Kant have problematized this idea: our knowledge of the world is always mediated, by the categories of our understanding according to Kant, or by language, anonymous structures or ideology according to more recent authors. How, then, can one make room for things? How can we ever know what things really are or want, when we are buried under representations, social constructions, ideologies or power relations? In opposition to the 'prelinguistic stance' of earlier thinkers, where knowledge of the thing in itself seemed possible, we are children of the linguistic turn: nothing is known without mediation through language.

One way to cope with things is to completely deny the truth of this linguistic turn (as some speculative realists have done). One could return to a prelinguistic position and claim that, at least for the sciences, a direct contact with the world is possible. As we saw in Chapter 4, this is not the option Michel Serres chooses, and neither do Bruno Latour and Isabelle Stengers. The Anthropocene as an epoch demands a different response, for it both shows that the traditional conceptions of science and technology cannot be maintained, and that purely linguistic approaches are unable to conceptualize ecological problems, since they cannot conceptualize the non-linguistic intrusion of nature in our politics. The science of climatology offers us a picture completely different from that of a science of certainty by direct contact with the facts themselves or a complete technological control. Rather, 'the very notion of objectivity has been totally subverted by the presence of humans in the phenomena to be described – and in the politics of tackling them' (Latour 2014, 2).

In the previous chapters we saw the necessary elements for this alternative. We have developed the distinction between the model of purification and the model of proliferation (Chapters 1–3); we have explored a relational ontology that blurs the boundary between thing and text (Chapter 4); and saw how an ecology of quasi-objects followed from it (Chapters 5–7). In this chapter I want to explore its value for an analysis of the Anthropocene. Specifically, I want to argue that Serres's anthropology of science can inspire such a different conception of technology, suited to deal with our current ecological crisis.

The theme of ecology plays a prominent role in the work of Michel Serres, which he himself recognized: 'I was one of the first, if not the first, to make of ecology, not just into a matter of fundamental urgency, but also a philosophical and even metaphysical question: the most ancient concept of philosophical thought, the idea of nature, had to be revaluated' (Serres 2014, 62). This is also how Serres is often appreciated in the secondary literature (e.g. Harris 1997; Watkin 2015). As Christopher Watkin rightly notes: 'Among the scandalously sparse secondary literature on Serres in general, ecology is one of the themes that has received a comparatively thorough treatment' (Watkins 2020, 332). According to Watkin, indeed, 'ecology is one of the areas in which his thought is prescient, even prophetic, and he anticipates the early twenty-first-century resurgence in ecological thought by over a decade' (Watkin 2020, 329).

It is no accident that Latour often refers to Michel Serres's *Le contrat naturel* (Latour 1999a, 2014, 2015a). In this book Serres was already trying to cope with the new problem of climate change: How can we deal with an active nature that refuses to play the role of inert matter, in which 'the earth is moved' by our actions (Serres 1990, 86)? Latour can be seen as a student of Serres, in the sense that both aim to develop a *postlinguistic stance*: interactions with things are still real and meaningful even if everything is mediated through language. Their philosophy is a postlinguistic philosophy aimed to let things speak again. As we saw in Chapter 4, it is through inscription devices that a connection between matter and text is established, but in the form of a negotiation. Although initially developed for different purposes, they see the Anthropocene and the 'intrusion of Gaia' (Stengers 2015, 44; Latour 2015a), or Biogea in the case of Serres (2010), as the ultimate proof that this postlinguistic correction of our views on nature, science and technology is necessary.

But Serres's philosophy should not be reduced to ecology. Ecology is neither the narrow focus of Serres nor an external addition or application of his thought.

One could argue that the core of Serres's thought is ecology, but not in the narrow sense as it is often understood. The same applies to the book *Le contrat naturel*. Though seemingly about ecology, this book never mentions the term, let alone give it a central spot. 'I carefully avoided the term' (Serres 2014, 233). Instead, Serres describes the book as a contribution to the philosophy of law and the question whether, and by what right, nature can be a subject of law. Simultaneously, the book aims to problematize any neat distinction between nature and society, since laws and contracts are found both in the natural and social realm.

Watkin suggests that one should make a distinction between a 'restricted ecology' and a 'general ecology'. Whereas the first 'reinforces the supposed dichotomy between a thoroughly human politics and a wild or unkempt nature, or between exclusively human environmental damage and an unspoiled world', the second, 'by contrast, seeks to find links, dependencies and passages between all the entities in a given milieu, travelling across dichotomies and back again' (Watkin 2020, 331–2). We find a very similar message in Latour's *Politiques de la nature* (1999a), where he pleads that the traditional concept of nature – that splits nature and society – forms in fact an obstacle to any decent political ecology. Thus, instead of an ecology in the narrow sense, focused on the preservation of a given and separate notion of nature, the general ecology of Serres is better labelled an *ecology of relations*, or even an *ecology of quasi-objects*: the political project of cultivating the right kind of relations to constitute a common world in which we can live.[2] As we will see, Serres often frames this ambition in terms of a shift from a parasitic relation to the world to one of symbiosis. We will therefore start with these two figures. Next I consider how, similar to Serres, Latour and Stengers mobilize their background in the sciences to look for this new postlinguistic perspective and the different view on technology it implies. Finally, I will re-examine the concepts of WAFEL, the parliament of things and the Anthropocene.

From parasite to symbiont

In Chapter 4 we saw how, for Serres, an analysis of communication implies that, instead of putting the struggle between messenger and receiver at the centre, one has to focus on the relation between them. This relation is not taken for granted, but has to be permanently constructed and maintained. This is done by the mutual war that messenger and receiver wage against a common enemy:

the background noise that must be silenced in order to communicate at all. Or as Serres puts it: 'To hold a dialogue is to suppose a third man and to seek to exclude him; a successful communication is the exclusion of the third man' (Serres 1969, 41). To create a relation you always have to invoke or exclude a third instance, the medium, that guarantees this. Think, for example, about the necessary silence of not only other people but also the world outside, to make someone able to read a text.

As we saw, this excluded third (*le tiers exclu*) is a central figure in the philosophy of Serres. It is present not only in linguistic communication but in every possible relation, including technological relations. To understand how relations are being destroyed or distorted on the one hand or amplified and created on the other hand, you need to focus on this third figure. Serres will respectively use the figures of the 'parasite' and the 'quasi-object'. The alternative to such parasitic relations, the natural contract, is one that instead starts from symbiosis.

The logic of the parasite

For Serres, the parasite is by definition always present because noise is always present (see Brown 2002). If the starting point is the world as one big relational network, then communication is not the establishment of relations, but the exclusions of the irrelevant ones. Not order but disorder is the starting point: 'The rational is a rare island which emerges, from time to time' (Serres 1977, 11). Order and communication, on the other hand, always have to be produced and made from this disorder. This is done, as we saw, through the act of 'translation': noise and interference are silenced, and incomprehensible clatter is translated into a common language that both messenger and receiver can understand.

This however leads to distortion in two ways. First of all, every creation of order implies a reduction, distortion and translation of this original communicative network. Yet the complexity of the network always exceeds the rational model, which can be applied to it (Serres 1982a, 174). In this sense, this implies a form of violence against a reality that is more complex than the models we use to talk about it. Secondly, a perfect exclusion is never possible, because not all parasites can be excluded. Practically, this is impossible because there are simply too many parasites and it is difficult to know which relations are essential and which are redundant. Logically, one ends up in a regression as well, because the act of exclusion is itself the creation of a new relation (and thus an invitation of new parasites).

Parasites are the noise that should be excluded if one wishes to communicate at all, but they can never be excluded completely. They can only be reduced to an acceptable level. In such a case, there will still be parasites but it will be claimed that they do not distort the message *in a relevant way*. This is however a mere claim, and they might still change the message in a relevant way without us being aware of it. By definition, a parasite will try to stay unnoticed by presenting itself as only transmitting the message without any distortion. Every communication, every technology and even every relation in general is thus open to this ambiguity. We already encountered this paradoxical logic when we spoke of Serres's view on the role of violence – and how any attempt to eradicate violence necessarily implies new forms of violence.

Although Serres mainly starts from examples about communication, this perspective can be applied to technology. For instance, this is the case when Latour uses Serres's ideas to make the distinction between intermediaries and mediators (Latour 2005b, 37–42). While we often think of technical instruments as unproblematic intermediaries, which transport a force without any distortion in a perfectly transparent way, we forget that the smoothness of this translation has to be produced. This is exactly the difficulty of technical interventions concerning climate change. The ideal is changing one thing for the better and keeping the rest stable. However, often unforeseen consequences will pop up. The smoothness of a certain technology is never a given fact, but the product of the mediator's optimization.

Every intermediary is an imperfectly disciplined mediator. Again, practically, one can assume that there will always be imperfections, noise. Logically, one can point to the paradoxical nature of every relation: 'If the relation succeeds, if it is perfect, optimum, and immediate; it disappears as a relation. If it is there, if it exists, that means that it failed. It is only mediation. Relation is nonrelation. And that is what the parasite is. . . . The best relation would be no relation. By definition it does not exist; if it exists, it is not observable' (Serres 1980a, 79).

To make this less abstract, let us return to Latour's study on Louis Pasteur (Latour 1984). As we saw in Chapter 4, the book's starting point is the relationship between people and their daily routines. These relations, however, can be disturbed by the noise of diseases, which parasitizes on human interactions, but often destroy them as well. Subsequently, physicians such as Louis Pasteur will present themselves as a way to 'smooth communication', namely by eliminating the germs through vaccination and pasteurization so that people can get back to their daily affairs. Although presented as such, this will not result in interactions free from any interference, but only in the exchange of one parasite (the germs) for another (the doctors). The doctors will introduce new distortions in human interactions, such as rules of hygiene or visits to the hospital. These are also alterations of the daily affairs, although not recognized as harmful.

We thus return to the logic of sacrificial violence we saw in Chapter 6. The creation of order always presupposes certain parasitic power relationships and violence (Serres 1977a, 12). In this sense, the replacement of one parasite for another does not necessarily imply better communication or a more stable technology, but can serve the surviving parasite itself. If it can convince the messenger and the receiver that it is the optimal medium, it will survive. Inspired by this, Serres has a great distrust of order, representation, language and consciousness, for they all are potentially driven by such parasitic power relations. They mutilate the noise of the world for their own survival, not for the greater good. Science is often reduced to a mere tool for hiding parasites beneath promises of smoother communication. 'Power wants order, knowledge offers it' (Serres 1977a, 12). Serres himself wants to get back to the things themselves: to give room to their multiplicity and their noise. The task of the philosopher is to protect this multiplicity, this inherent potential of all things (Serres 1980a, 46). This means that a philosopher should be the voice of this forgotten disorder beneath all constructed order.

Serres has the ambition to go beyond language and culture to the things themselves, because language is merely an imposed order on the multiplicity of things. 'Can we step outside our language' (Serres 1985, 89)? Serres wants to restore the speech of things. However, it is not the case that things are silent, despite having the potential to speak. Things do always already speak, they always emit noise and thus potential information.

Here the excluded third is seen as a positive figure: that is, noise that always breaks through our cages of language: 'the third person provides a foundation for the whole of the external real, for objectivity in its totality, unique and universal, outside any first- or second-person subject' (Serres 1991a, 48). However, this noise of the world is not recognized. This is the real issue for Serres: somehow everything always speaks, but we ignore this fact. As we will see, this is also the case in the Anthropocene: we silenced nature and forced it to play a passive role without ever realizing the violence our relations inflict on the world. Serres attempts to go back to this moment of noise, before things are being silenced, by again falling back on the concept of the 'quasi-object'.

The omnipresence of the quasi-object

We already encountered how, for Serres, a quasi-object is something that predates the subject-object distinction. It is neither an active subject nor a passive object, but instead the ground for both of them. It creates a network around itself that makes agency and structure possible. The most famous example he gives is that of the ball within a game. The ball is not a passive object, but the whole game moves around it, and even creates the collective:

> Let us consider the one who holds [the ball]. If he makes it move around him, he is awkward, a bad player. The ball isn't there for the body; the exact contrary is true: the body is the object of the ball; the subject moves around this sun. Skill with the ball is recognized in the player who follows the ball and serves it instead of making it follow him and using it. . . . Playing is nothing else but making oneself the attribute of the ball as a substance. The laws are written for it, defined relative to it, and we bend to these laws. Skill with the ball supposes a Ptolemaic revolution of which few theoreticians are capable, since they are accustomed to being subjects in a Copernican world where objects are slaves. (Serres 1980a, 226)

This quasi-object must not necessarily be an 'object', such as a ball or a piece of money (Serres 1982a, 148–9), but can also be a quasi-subject: a leader, a king, a celebrity (Serres 1987, 181–2). We already encountered in Chapter 6 how Serres uses Dumézil's tripartite system to map three common types of quasi-objects: a fetish (Jupiter), a weapon (Mars) and a commodity (Quirinus). However, and this is crucial, the quasi-object is nothing without its relations to the things around it, its conditions of possibility. Its existence depends on the things around itself, it is itself nothing more than a node of these relations. In this sense, it is an object of which the relations to other things and persons cannot be forgotten; or a subject of which the necessity of the things around it to make him or her speak, move or think is recognized. 'A ball is not an ordinary object, for it is what it is only if a subject holds it. Over there, on the ground, it is nothing; it is stupid' (Serres 1980a, 225). The ball is nothing without the players. The king is naked without its clothes.

These quasi-objects are the ground for the collective of subjects and objects, for our relationships. In fact, reality consists mainly of quasi-objects, rather than orderly subjects and objects, which are the exception. To illustrate this point, Serres uses the image of the discovery of irrational numbers (famously through the diagonal of a square with sides of length one). For Greek philosophers, there were only rational numbers, but this diagonal confronted them with a new world:

> From this contradiction, the third should have been excluded. But if that were the case, the said diagonal wouldn't exist; . . . From then on, the discovery of real numbers, spurting like a geyser from this absent fault line, insists that

all other known numbers, at least in those days, be reduced to limit cases of this new form. . . . Soon one will not find anything but this third, as soon as its exclusion is pronounced. It was nothing, see how it becomes everything-or almost. (Serres 1991a, 45)

In the same way, quasi-objects will become the rule, rather the exception once we recognize their existence. Serres's ambition is to give them a rightful place in the political scene. As we will see, the Anthropocene holds the promise to give them proper political representation. But, to get to the politics of quasi-objects, we first need to understand (a) what kind of technological relations are possible with quasi-objects and (b) why our modern approach ignored quasi-objects in the first place. For this, Latour's and Stengers's philosophy of science is helpful.

Technology as negotiation

In the case of both Serres and Latour, their early work was grounded in a philosophy of science. In the case of Serres it was an examination of the new new scientific spirit; for Latour, it was a study of the production of scientific facts within a laboratory. In both cases, this will form the basis for an analysis of current ecological crisis.

In his early work, Latour finds a general distinction between *ready-made science* and *science in the making*. As we saw (see Chapter 2), Latour will describe this duality as the *translation* and the *purification* of quasi-objects (Latour 1991, 11). According to Latour, the reason why modern science is so successful depends on its abilities (a) to recruit and connect a high number of relevant allies, both humans and non-humans, who will affirm the theory and (b) to make this whole construction and recruitment process invisible as if one was merely describing a passive nature (Latour 1987a, 106; 1991, 108).

Although again mainly concerned with science, the role of technology is crucial here because for Latour science often, if not always, boils down to technoscience. However, in this modern view of science, technology is approached in a very specific way. For instance, in (b) the role of technical instruments is reduced to a mere purification of facts that were already present beforehand, waiting to be discovered, while in practice we are faced with numerous quasi-objects. On the other hand, (a) echoes Serres's idea that all relations imply parasites and that mediators have to be translated into intermediaries. The art of science seems to lie in its practices to successfully translate phenomena into scientific facts, without creating any relevant distortions. It will be able to claim that the scientist has not been a parasite, but only 'smoothed the communication' between object and subject, although, in practice, the scientist has distorted the phenomena in some way. For this, technology is crucial, but a specific technology whose role is not recognized from the moment the translation is finished.

From a quasi-object to a witness

To understand the precise role of technology in this model, the work of Stengers can be very helpful. For her, every scientific claim starts as a *fiction*: that is, a claim about reality that does not distinguish itself from other claims about that same reality. This is what the linguistic turn implies: every claim is always open to the accusation of being merely a representation. 'Normally, any phenomenon that we observe can 'be saved' in multiple ways, each way referring to a human author, his projects, his convictions, and his whims' (Stengers 1997, 156). However, the construction of scientific facts implies the abnormal case, which is the creation of a *difference*, a *non-equivalence*. The scientist has to construct a case in which she can claim that she is not speaking in her own name, but in the name of things, in the name of nature. This is being done by introducing technological interventions in the phenomena one is studying. However, the scientist also has to make her own mediation between nature and our understanding as invisible as possible, and thus, as Serres remarked, make the relation itself disappear. 'What does matter is that [her] colleagues be constrained to recognize that they cannot turn this title of author into an argument against [her], that they cannot localize the flaw that would allow them to affirm that the one who claims "to have made nature speak" has in fact spoken in its place' (Stengers 1997, 160).

We saw in the previous chapter that, by constructing a reliable *witness*, a phenomenon can testify for the theory, introducing a form of transcendence into the scientific discourse. The scientist, thus, has 'to produce a testimony that cannot be disqualified by being attributed to her own "subjectivity", to his biased reading, a testimony that others must accept, a testimony for which he or she will be recognized as a faithful representative and that will not betray him or her to the first colleague who comes along' (Stengers 1997, 88). She has to present herself as the perfect parasite that merely transports reality to our discussions. The scientist transcends the mere linguistic stance by constructing a third party, using technological interventions, who will be recognized as a reliable witness. It vouches for her, that she does not distort reality, but translates its information without transformation. She has to succeed in making one admit that the reality [s]he has fabricated is capable of supporting a faithful witness, that is to say, that [her] fabrication can claim the title of a simple purification, an elimination of parasites, a practical staging of the categories with which it is legitimate to interrogate the object. The artifact must be recognized as being irreducible to an artifact'. (Stengers 1993, 167)

The role of these technical instruments and laboratories is that they constitute this difference, they 'discipline' the quasi-object, the phenomena, to be the best possible witness, which only affirms the theory of the scientist and thus appears as a mere passive but affirming object. The third is excluded, noise becomes information; it will only say one thing and nothing else. This is, according to Stengers, the core of the modern experimental practice: *'the invention of the power to confer on things the power of conferring on the experimenter the power to speak in their name'* (Emphasis in original) (Stengers 1997, 165). The core of the scientific practice lies in its technical potential to translate ambiguous noise into reliable witnesses.

How to negotiate with things

However, this does not imply that scientists merely impose their will on the phenomena. This is not a submission of objects (Stengers 2013, 189–90). This leads us back to the claim of Serres. He claimed that all representations are problematic, but this might be a step too far. Serres is ambiguous whether all order is violence that should be avoided or that some specific forms of order, namely those in function of specific power relations, are the real problem. We can refer back to a strategy of Serres we already encountered in the first chapter, and that Watkin (2020) highlights: *opposition through generalization*. Serres expresses his disagreement with a specific way of how quasi-objects are disciplined, not by criticizing this organization, but by showing that it is merely one instance of an infinite set of possibilities. 'It is a mode of engagement that does not critique or deconstruct, but nevertheless ends by destabilising' (Watkin 2020, 89). We can think back to the example of the leprechaun.

Latour and Stengers, however, are more inclined to claim that some constructions are acceptable while others are not, because the submission of things is not the whole story. The impression of submission is only created after the whole construction process is over. During the construction of these facts, technology plays a different role, namely that of the order of *negotiation*: one has to listen to the quasi-objects and their relations and try to persuade them to follow your theory, while at the same time you are being persuaded by them. To get to the reliable witnesses, the scientist has to go through a process of carefully and accurately putting her research object into question, and at the same time *being put into question by it*. For Stengers, good science is able to put itself *at risk*: that is, to give the object in question the power to put the subjectivity of the scientist and her categories into question (Stengers 1997, 126; 2000, 134). Bad science, on the other hand, is defined by Stengers as the mutilation or forgetting the object by a science. So bad science starts with passive objects, rather than ends with them. But during genuine negotiations the phenomenon can dismiss the scientist's questions as irrelevant. Things can respond and show themselves to disagree with the questions asked; they too can take the lead, similar to the quasi-objects of Serres. If this is recognized, good scientific facts can be constructed.

Stengers herself illustrates this by a discussion between Diderot and D'Alembert: while D'Alembert is a follower of a very rigid form of mechanic materialism inspired by Newton, Diderot presents him with the case of an egg, a complex chemical-biological entity. In this case the materialism of Diderot is described by Stengers as a *demanding* materialism and not a reductionist one: 'What Diderot asks D'Alembert is that he *give* to the egg the *power to challenge* his well-defined categories' (Stengers 2011b, 373). A good scientist will let the egg be a *risk* which can challenge her own ideas of materialism. Another example is given by Vinciane Despret (2015), who is strongly influenced by Stengers. At a certain moment behaviourists tried to use the skinner box on a different organism than the eternal pigeon, namely the raven. The raven, however, refused to pull the levers and instead destroyed the box. The disappointed behaviourists returned to studying pigeons, but, with this decision, started to do bad science. Instead, they could have seen the reply of the raven as a lesson that the technology and categories they used to study organisms are inadequate and should be changed. Good science would in this case be open to the answer of the raven.

From this perspective, another use of technology comes forward, namely one based on negotiation. Good science can only occur if the proper technologies are in place. The role of technology here is not to enforce itself on the phenomena and reduce them to passive objects. Rather, technologies are being mobilized to allow the phenomena to articulate themselves as quasi-objects, by being sensitive to its feedback.

The two faces of science thus imply two roles of technology: on the one hand technologies are introduced to create the possibilities to listen to things, to create a feedback loop between the scientist and the quasi-object, both posing their own questions. The power of the sciences, to go beyond the linguistic turn, consists of their ability to incorporate *both humans and non-humans* into their networks. On the other hand there is the purification of the quasi-object once the negotiation is over: it is stabilized into a witness with predictable behaviour. Science will present itself as if its rational subjects spoke in the name of a silent nature. This is what Latour calls our modern condition: quasi-objects are forgotten and reduced to active subjects and passive objects (Latour 1991, 139).

What is a parliament of things?

If this purification and translation is the essence of science, and it seems to work, then what is wrong with disciplining quasi-objects into objects? Why do these quasi-objects need to be heard as quasi-objects? As stated earlier, the problem is not that quasi-objects are being disciplined into objects per se, but rather that this is being done in a problematic way. This is often the ground of a misunderstanding. Yves Gingras, for example, criticizes Latour's perspective because 'it is impossible to write or even think without making distinctions' (Gingras 1995, 125). In a similar vein Latour has been criticized for first claiming that all subjects and objects are constructions but then 'glibly employ[ing] all such [objects and subjects] without going into ontological *Angst*' (Zammito 2004, 201).

Latour's aim is not to bring us back into an endless limbo of letting quasi-objects speak in their multiplicity, in contrast with what Serres's view might suggest. In his concept of the 'parliament of things' the stress is too often placed on *things* while ignoring the *parliament*. The goal is not to end up in a world with no objects or subjects, but rather in one where there are only well-constructed objects and subjects: that is, objects and subjects that are the result of adequate negotiations between all relevant actors involved in the network. Latour is thus in favour of, eventually, picking out one organization as superior. But for this we need to re-evaluate our institutions, and therefore we need to construct and adequate *parliament* of things.

It is therefore not a plea for less technology or less reduction, but rather for more technology and more reduction, yet thoughtful use of technology and deliberative reduction. Or as Latour states: 'The moderns were not mistaken in seeking objective nonhumans and free societies. They were mistaken only in their certainty that that double production required an absolute distinction between the two terms and the continual repression of the work of mediation' (Latour 1991, 140). Similarly, Serres is not pleading in his work to give up any attempt to control or shape the earth, to abandon technology. Instead, he pleads for 'more mastery', in the sense that we not only have to master the world but also our own mastery of that world. We already saw how we are now in a situation where we depend on things that depend on us. We therefore have the task not only to control and manage the dependence of things, but also our dependence on these dependences. Or as Serres phrases it in an interview with Latour: 'We have resolved the Cartesian question: "How can we dominate the world?" Will we know how to resolve the next one: "How can we dominate our domination; how can we master our own mastery"' (Serres and Latour 1992, 172)?

Or, since quasi-objects cooperate in networks that support our current collective, the problem is that the *current* composition of our collective is inadequate. According to Serres and Latour, the main issue is that this purification of quasi-objects is (a) not always successful and (b) has become more problematic in the age of the Anthropocene.

First, it is not always successful because quasi-objects can be adequately transformed into objects within the laboratory settings, but these settings have their limits. This seems to be a central message of Serres's focus on chaotic object such as whirlpools (see Serres 1977b). Similarly, as Stengers argues, not all sciences succeed in creating such legitimate witnesses out of quasi-objects, even if they claim to do so (Stengers 1997, 88). Some sciences, such as biology or political science, would do better 'to follow' their study objects and recognize them as quasi-objects (Stengers 1993, 144–5). The experimental sciences should be seen as an exceptional event, rather than the rule. By taking the purified object of laboratory physics as the paradigm, one is unable to understand what is going in within more complex fields such as biology or economics. By being aware that we are initially always dealing with quasi-objects, we can recognize, as for instance ethology argues, the blind spots that follow from manipulating birds only in a very one-sidedly purified skinner box (see Despret 2015). This can finally allow us to work to an adequate purification of these objects, that take all relations of the quasi-object into account, or even recognize that we are unable to purify certain quasi-objects we are faced with.

Secondly, it has also become more problematic to ignore quasi-objects due to global problems such as the ecological crisis (Serres 1990; Latour 1999a). As Latour states, the class of quasi-objects 'ends up being too numerous to feel that it is faithfully represented either by the order of objects or by the order of subjects' (Latour 1991, 49). Latour describes this event paradoxically as the 'end of nature'. According to him, we always have had an idea of nature, but saw it as something merely passively out there without any agency – a pure collection of means to our ends. Now, however, such an idea has become untenable: 'the repressed has returned' (Latour 1991, 77). Nature does seem to respond, react and reply in many unpredictable but devastating ways to our behaviour. Our collective is inadequate, in the sense that the current proposed purifications of quasi-objects are unable to assign a place to all relations and actions of the objects. The unaccounted actions of the objects therefore build up, until finally they are too immense to ignore and in fact threaten the stability of the rest of the collective. The objects show themselves as quasi-objects again. We enter a 'crisis of objectivity': 'Political ecology thus does not reveal itself owing to a crisis of ecological objects, but through a generalized constitutional crisis that bears upon *all objects*' (Latour 1999a, 20).

Latour repeats and radicalizes this message in his recent work on the Anthropocene, for instance in his book *Face à Gaïa*. For Latour the term 'Anthropocene' shows what a notion such as 'ecological crisis' did not, namely that it is not a temporary state that will pass by. '[T]hat which would possibly be nothing but a passing crisis is being transformed in a profound alteration of our relation to the world' (Latour 2015a, 17). While in earlier work Latour was mainly describing how we have never been modern, he uses the notion of the Anthropocene to define us in an affirmative way. The new condition in the Anthropocene implies several things for Latour.

First of all, for Latour the Anthropocene does not imply some kind of radical break, a fundamental revolution in earth's history. We do not live in another world, but the Anthropocene implies first and foremost the obligation to relate to the old world in a fundamentally different way. In this new view, the traditional players disappear (nature, humanity, science, technology) or rather we realize that they have never existed in the first place.

Secondly, nature has ceased to play the passive role that our modern science and technology forced upon it. Latour and Stengers rather speak of 'Gaia', a term introduced by James Lovelock in 1969. But Lovelock, according to Latour, is often misunderstood. Gaia does not mean that the earth has become a living organism or that it is a fixed and closed system, but rather that it must be seen as 'the name proposed for all the interwoven and unpredictable consequences of the acting powers, each of which pursues its own interest in manipulating its own environment' (Latour 2015a, 187). Gaia has no fixed identity nor can serve as a transcendent judge of our conflicts, showing the objectivity behind our subjective claims. Rather she is the third in the sense of which Serres speaks, namely of a parasite or quasi-object constantly intervening and changing our relations. 'Gaia is a third party in all our conflicts – especially since the

Anthropocene – but she never plays the role of third party *superior* to situations and able to *command* them' (Latour 2015a, 307).

Serres himself avoids the term 'Gaia' and is in fact very critical of the term. When asked in an interview whether his own reflections on the Earth as a Grand Fetish are not similar to Latour's notion of Gaia, Serres responded as followed:

> The Gaia hypothesis is a new age hypothesis. I don't like it for a simple reason: a living being reproduces itself – this is one of the conditions of life. However, it is quite obvious that the planet does not reproduce itself. Therefore, to say that Gaia is alive is untenable for a scientific mind. I recognize of course that the Earth, considered as a whole, has certain characteristics of life, in particular self-regulation, but it stops there. The Earth does not evolve in the Darwinian sense of the term. (Serres 2014, 266)

Though a common critique, we saw how Latour is quite explicit is dissociating himself from such a claim. In that sense, as Watkin also notes, 'though Serres's criticism of Gaia strikes true for Lovelock, it rings hollow for Latour' (Watkin 2020, 368). Serres uses his own alternative term, namely that of *Biogea*: '"Bio" is life, and 'Gea", in Greek, is Earth. Biogea is our habitat, what we live in, our world' (Serres 2014, 248). It is striking how similar Serres's talk on Biogea is that to Latour's use of Gaia. To cite just one passage:

> Ice melting, waters rising, hurricanes, infectious pandemic diseases: Biogea is starting to scream. The global world, although stable beneath our feet, is suddenly falling on the heads of women and men. They had expected it so little that they are wondering how in their world-less society to deal with the sciences that were turned to the things of the world, and have added up and measured its sovereign forces and heard strange voices. Panic time, the Great Pan is back! (Serres 2009, 42)

A third implication of the new condition of the Anthropocene is that it implies the end not only of nature but also that of humans. The idea that humanity has become the most influential factor in the Anthropocene is thus easily misunderstood, according to Latour. For one, it is misunderstood on an empirical level because it is misleading to speak about humanity, as if we are now faced with a unified collective. In fact, no such unification exists and not all of humanity is equally responsible for the Anthropocene: native tribes in the Amazonian rainforests are not influencing the planet in the same way as Western industries are. 'The Anthropos of the Anthropocene? That is Babel *after* the fall of the giant tower' (Latour 2015a, 189).

But the Anthropocene is also misunderstood on a more conceptual level since the opposition does not consist of active humans versus passive nature anymore. The traditional picture of the human is that of the subject: that is, someone who possesses all the agency and the unbounded capacity to do with the planet what (s)he wants. But being a subject in this age 'is not acting autonomously in relation to an objective framework but sharing the power to act with other subjects who also lose their autonomy' (Latour 2015a, 84). To influence means not that one has all the power over something, but rather that all those being influenced have part of the agency, namely to react and to respond. The Anthropocene forces us to become full-blown quasi-subjects again, and thus depend on the sensitive networks of Gaia. We already encountered similar thoughts in Serres reflections on the Earth as a Grand Fetish.

Finally, the players of science and technology change also. Science cannot be an ultimate referee anymore, but must recognize its dependence on its own networks. We enter an era, the Anthropocene, where science is never certain, but where uncertainty has become one of its main characteristics.[3] Technology, in a similar vein, cannot see itself as a tool in the hand of the active subject, but must play a quite different role in the negotiation with the quasi-objects, with Gaia, as we will see in the next section.

As already indicated, Latour is inspired by the book *Le contrat naturel* by Serres, in which he claimed that our modern social contract is insufficient, because it always excluded things although we necessarily relate to them. Within the Anthropocene, we are faced with the '*generalized revolts of the means*: no entity – whale, river, climate, earthworm, tree, calf, cow, pig, brood – agrees any longer to be treated "simply as a means" but insists on being treated "always also as an end"' (Latour 1999a, 155–6). The Anthropocene is nothing more than a general intrusion of quasi-objects in our current political collective: it demands us to reopen the negotiations with the quasi-objects we relate to, for otherwise the world will change and opt for a new political collective without humans.

We are in need of a peace treaty, or what Serres calls, a natural contract. Serres often phrases this alternative in terms of *symbiosis,* in opposition to the parasite. 'The natural world will never again be our property, either private or common, but our symbiont' (Serres 1990, 44). Instead of the parasitic relation we now have towards the Earth and its objects – we take without paying our dues, without acknowledgement – we need to constitute a world in which all the actors can accept the terms of interaction. Why? First of all, out of ethical reasons: all actors deserve to be treated as more than a mere means to our ends, but as a goal in itself. Latour, quite provocatively, takes up the famous Kantian imperative not to treat human beings simply as a means but always also as ends, but adds 'that we *extend it to nonhumans as well*' (Latour 1999a, 155).

However, there is a second set of reasons to do so, namely of a practical nature: though a parasitic relation can be beneficial for one party on the short term, we have reached a point where we, the parasites, risk to kill our host – and no other host is available. 'Intelligent, efficient, the parasite wins its battles day after day, but it always ends up losing the war, when, precisely, its hosts are exhausted. Henceforth, but I have said it enough, we must treat nature as a symbiote and no longer as a treasure' (Serres 2006, 138). Hence, the natural contract can be understood as a plea for symbiosis:

> Back to nature then! That means we must add to the exclusively social contract a natural contract of symbiosis and reciprocity in which our relationship to things would set aside mastery and possession in favour of admiring attention, reciprocity, contemplation, and respect; where knowledge would no longer imply property, nor action mastery, nor would property and mastery imply their excremental results and origins. An armistice contract in the objective war, a contract of symbiosis, for a symbiote recognises the host's rights, whereas a parasite – which is what we are now – condemns to death the one he pillages and inhabits, not realising that in the long run he is condemning himself to death too. (Serres 1990, 38)

This means a contract in which quasi-objects are recognized. But how can we achieve this?

The counterclaims of quasi-objects

As stated earlier, both the prelinguistic and linguistic positions seem implausible. Here we have to take a postlinguistic stance: we should open up a space so things can give feedback to our current collective in order to reopen negotiations and propose reforms for a better political collective. This is possible because the world is always connected with us, since non-humans always emit noise, and thus possible information (see Chapter 4). Or as Serres states: 'In fact, the Earth speaks to us in terms of forces, bonds, and interactions, and that's enough to make a contract' (Serres 1990, 39).

However, we tend to ignore these forces. One way in which Serres diagnoses this problem is in terms of *acosmism* (Serres 1977, 288; 2006, 137): we forget or ignore the world we live in, the relations that are the necessary conditions for us to live in. We only have eye for culture, politics and society. There is room only for a social contract, which ignores nature and objects, which are nonetheless its conditions of possibility. Serres links this forgetting with several factors, which we encountered in previous chapters: the disappearance of agriculture, the new biotechnology and the current digital revolution. Nonetheless, as previous chapters have argued, we are defined and determined by our environment, our relations, our quasi-objects. And paradoxically, our current ecological crisis highlights the unsustainability of this stance:

> In philosophy, in pedagogy, but above all in politics, acosmism is now becoming dangerous for men themselves. To continue to live by thinking of themselves alone in the world, the world itself risks taking away all their life. Their actions as subjects transform them into passive objects, suddenly subjected to the forces of a world which, according to the loop already described, becomes the subject of this action in return. Hence the strange Natural Contract that I proposed, little understood, although quite readable. (Serres 2001, 253)

To understand this paradox we can invoke again one of the radical hypotheses of Latour: it is precisely this blindness towards quasi-objects that makes our modern technologies possible. Or in the words of Latour: 'the more we forbid ourselves to conceive of hybrids, the more possible their interbreeding becomes' (Latour 1991, 12). To transform our bodies and our world to such an extend requires a certain negligence towards how we are defined by them and an insensitivity towards the impact of such transformations on the world and ourselves.

To become more sensitive to the noise of the world we need different technologies. Otherwise the quasi-objects would keep revolting against our current collective, which is exclusively focused on humans. This is what Serres (1990) means when he calls for a natural contract and Latour in his response of a 'parliament of things' (Latour 1991, 142; 1999a): to let things re-enter politics by allowing them room to articulate their habits, behaviours and claims. We thus return to the question of an ecology of quasi-objects: Which relations should we cultivate and which should we avoid? Serres tries to conceptualize the answer to this in terms of a plea for a *cosmocracy* (2008a, b): a kind of politics that takes up this question of relations, of how we are defined by the world around us.

How is this possible? Well, to get a grasp on how they would be able to articulate their claims, in fact we can learn a lot from the sciences, not *ready-made science*, but *science in the making*. We need to return to the position of the scientist-at-risk: that is, someone who is aware she is dealing with a quasi-object that can speak, and put her own categories into question. The Anthropocene requires a specific type of technology, namely not the *technologies of control*, but rather *technologies of negotiation*: technologies which enable us to become more sensitive to the reactions and relations of the quasi-objects. Technology in the Anthropocene is not only about trying to control environmental factors such as rising temperatures but first and foremost has the goal to detect the way these quasi-objects respond to our actions. We make constant claims about how Gaia will respond to our policies, for instance, by claiming the CO_2 level will decrease by a certain policy. Technologies of negotiation are there to give quasi-objects such as CO_2 room to respond. Following our policy CO_2 will either accept our claim (by decreasing) or will make a *counterclaim* (by doing something else, such as increasing). Of course, the politician or the scientist can make a new claim to cope with this resistance, but the ideal (both in political democracy and in scientific practice) is that *a feedback mechanism* is respected: those who are represented should always be able to make a counterclaim and be heard.

In this manner we must understand the aim of Serres's proposal for WAFEL, Latour's parliament of things (1999a) or Stengers's *cosmopolitics* (2005, 2010, 2011a): do not define beforehand what things are, but offer them the opportunity to call your own questions and perspectives into question. We can define what the objects are only after these proper negotiations, although some may never be well defined, such as GMO's or climate change. There is no guarantee that it will work, but neither that it will fail. Or as Latour states:

> The deliberations of the collective must no longer be suspended or short-circuited by some definitive knowledge, since nature no longer gives any right that would be contrary to the exercise of public life. The collective does not claim to know, but it has to experiment in such a way that it can learn in the course of the trial. Its entire normative capacity depends henceforth on the difference that it is going to be able to register between t_0 and $t + 1$ while entrusting its fate to the small transcendence of external realities. (Latour 1999a, 196)

This is precisely the role of technologies in the Anthropocene: making us as sensitive as possible to the differences and changes that occur when we introduce a certain intervention. In this sense, the parliament of things itself can be understood as a technology of the Anthropocene, namely one not aimed to help us to fully control nature, but first and foremost help us to negotiate to remain part of a collective with as many possible quasi-objects taken into account. This is a never-ending process, simply because there will always be elements that have not been taken into account yet. Precisely because our collective will always change, we need constant monitoring.

Two issues, however, remain. First of all there is the question of who would speak in name of all these quasi-objects. Serres's answer seems to be: scientists, who 'should speak in the name of things, in the language of the things themselves, to speak in the WAFEL' (Serres 2009, 51). An obvious reply to this proposal is the question how to assure that these scientists do not form an elite which only serve their own interests (see Moser 2016, 96–9).

To counter the possibility of such a perversion of interests, Serres argues that members of WAFEL should take an oath:

> For scholars to speak in the name of Biogea requires that they first take an oath whose terms must free them from any allegiance to the three traditional classes. To become credible they must, as *secular people, swear they will not serve any military or economic interest*. Only at this price can they speak in the name of Biogea at the WAFEL. (Emphasis in original) (Serres 2009, 65)

As some have noted, this is quite an 'underwhelming solution' (Watkin 2020, 376). Latour tries to tackle the same problem by the opposite road: by recognizing divergent interests, but in an explicit form. Divergent interests should be acknowledged, instead of being hidden behind a misleading idea of objectivity or the public good:

> it is no longer necessary to hide behind some appeal to the objectivity of knowledge, to the incontrovertible values of human development, to the Public Good or to the well-being of common humanity. Tell us, rather, who you are, who are your friends and your enemies, *whom* you are ready to sacrifice to your own happiness, *which* foreigners can put you in a situation such that your existence will be denied – and, in addition, please tell us clearly, finally, by *what deity* you feel convoked and protected. (Emphasis in original) (Latour 2015a, 247)

Watkin interestingly captures this two opposing models by reference to the difference between Anglo-Saxon democratic politics and its French Republican counterpart:

> Latour's parliament of things is an extension of representative democracy: each human and non-human 'concern' receives political representation in the parliament. Serres's emphasis, by contrast, is not on the communitarian notion of each interest group receiving its voice at the table but on the commonality of all the members of the cosmocracy. (Watkin 2020, 374)

Watkin himself also suggests a second way out of this problem, inspired by Stephanie Posthumus. Similarly to how we do not need to see the natural contract or the parliament of things as a prescription for a specific set of practical procedures, but instead as a 'metaphor' by which we can reconceptualize how we think about politics, we can see WAFEL as 'as a way of casting a vision that can open the possibility of change and cause people to see the world and its institutions differently, not as a point by point plan for implementation in the next five years' (Watkin 2020, 377).

Besides this first problem, a second issue also remains: Does this model imply that we have to live together with everything, regardless of how problematic or dangerous these entities might be? Is there still room for dissent, opposition and disagreement? Watkin raises this question explicitly, and tries to explore in what way Serres's point of view is not committed to a 'motherhood and apple pie response to the environmental crisis' (Watkin 2020, 371). He gives the example of racist ideologies: Should we just live together with racists or fascists, in a symbiotic relation?

A somewhat similar critique is raised by Stengers, who argues that the early Latour (1991) tended to be rather optimistic about the fact that all quasi-objects could become part of one parliament of things. The essential underlying assumption is that everything can be an object of negotiation, but this implies a form of exclusion, namely of those things that refuse to cooperate. 'No one can introduce themselves by establishing conditions – take it or leave it – from which the possibility or impossibility of agreement would follow' (Stengers 2011a, 347). However, the construction of certain elements cannot be the object of negotiation, even in such an open parliament of things. There are things such as gods, values and practices that we do not want to give up because they constitute our fundamental identity. Such claims are definitely not without ground. For instance, without the specific experimental practices of the scientists, they cannot do science; but similar claims can be made about religions or ethical values. In this sense, Stengers proposes a corrected 'Cosmopolitical Parliament' (Stengers 2011a, 395) where there is room for these excluded elements, that refuse to take part, but cannot be ignored.

From Serres's perspective, part of the answer lies in his take on universalism: a universalism that does not deny individuality and particularisms, but is universal in its variation (see Chapter 5). Similarly, the model of symbiosis does not commit us to saying that we all are or should be the same. As Stengers also notes, symbiosis differs from consensus. Consensus implies that all parties agree on the goal and the means to get there. Symbiosis, however, merely entails that all parties are interested in success, but can define their motivations and even success differently. Symbionts have references to one another, in opposition to parasites that ignore their hosts, but these references can differ radically from one another. Stengers (2011b, 379) refers to Deleuze and Guattari's (1980, 11) example of the wasp and the orchid: though both engage in a clear form of symbiosis, they do not share a common goal or definition of each other's motivations. Symbiosis thus entails the 'stability of a relation without reference to an interest that would transcend its terms' (Stengers 2010, 36).

What is required instead is a form of diplomacy where all parties come to an agreement of how to symbiotically live with one another – and this entails a renegotiation of their identities, relations: the things the actors are willing to give up and the other things they do not. Thus, as Watkin concludes, 'a symbiotic response to the curse of racism would be to identify and redirect the lust for affiliation and domination that lie at its heart into, respectively, cosmocratic belonging and the mastery of mastery' (Watkin 2020, 375). If some actors are unable to accommodate, a (temporary) exclusion from the cosmocracy might therefore be warranted. But there must always be a possibility for these excluded actors to represent their case, in the hope that a future symbiosis can be possible.

In his later work, these issues also made Latour revise his initial conception of the parliament of things. Negotiation is never by definition successful, and therefore one has to obtain the model of the diplomat, rather than the expert (see Latour 1999a, 209–17). Diplomacy means that one is fully aware that one is taking risks in the negotiations, that one does not have nature on one's side as *ready-made science* once claimed, nor that negotiations will always succeed. We cannot start from a given world, but we have the political task of 'the progressive composition of the common world' (Latour 1999a, 18). 'Diplomacy . . . celebrates another, quite artificial, conception of truth – what is true is what succeeds in producing a communication between diverging parties, without anything in common being discovered or advanced' (Stengers 2013, 194).

In a similar vein, this might explain some recent criticisms of Serres's original proposal, found in the work of both Latour and Stengers. For Latour, a natural contract has become impossible 'because in a quarter of a century, things have become so urgent and violent that the somewhat pacific project of a contract among parties seems unreachable. War is infinitely more likely than contract' (Latour 2014, 5). Related to that, Stengers talks about the 'intrusion of Gaia', which does not ask a response of us, but rather obliges us to find means to protect ourselves from this new condition (Stengers 2015). The necessity of negotiation, presupposed by Serres and the early Latour, is no guarantee anymore. We cannot be certain that we will find a new natural contract.

However, Latour and Stengers tend to make the stronger claim: such a contract has become impossible. Similar to the criticism of the necessity of noise or that of quasi-objects, however, one can state that this claim mixes up two different notions, namely the *certainty of the impossibility* of a contract and the *uncertainty of its possibility*. Their claims seem only to support the second claim and not the first claim. In the Anthropocene a natural contract might still be a possibility, although we cannot even be certain of that.

Conclusion

For Serres, Latour and Stengers the problems of the Anthropocene require new technologies to deal with it, including new institution, such as WAFEL or a parliament of things, which is neither a parliament of subjects nor a parliament of objects. Rather it aims to be a parliament of quasi-objects: that is, a place where both objects and subjects are represented, but together with their relations, their scientists, their uncertainties and so on. The problem was never that non-humans did not speak, but that they were forced to speak only in one way, namely as mere passive objects. Once again, the conclusion is not that we should just let the original chaos of the world speak. In the case of technology, the message is not to abandon all technologies since they imply this violence. Instead, the task is to listen to things as things, to create new technologies in which quasi-objects can express themselves in their complexity and multiplicity, to articulate and differentiate their habits and their associations (Latour 1999a). Otherwise, 'we shall remain barbarians besieged by inhumans – and before Gaia we shall remain without a voice' (Latour 2013a, 288).

Subsequently, we do not need to stop constructing objects, but rather construct them better by consciously involving the quasi-objects in the construction process, by constructing WAFEL, an adequate parliament of things. Nevertheless, as the critique of Stengers shows, there is no guarantee that such a parliament of things will work. But also, and contrary to the recent pessimism of Stengers and Latour, none that it will fail.

Notes

Chapter 1

1 This has to do with the institutional background of philosophy of science in France, as well as with particularities of the philosophy of Auguste Comte, who stressed the necessity of the history of science to understand the functioning of the rational mind (see Chimisso 2001; 2008a; Brenner 2015).
2 Serres also links Comte's three states to Dumézil's tripartite hypothesis of law-religion (Jupiter), war (Mars) and economy-agriculture (Quirinus) (see Chapter 6). Nevertheless, for Serres, 'Comte's diagram, which identifies three regimes (rex), is entirely ordered from the point of view of the first function, namely Jupiter' (Serres 1977a, 175). In contrast, Hegel reduces all to the second function of Mars (the dialectics of master and slave) and Marx places Quirinus at the centre (Serres 1977a, 176).
3 At other moments, however, he seems to turn the scheme around: 'We have moved from solid to liquid to gas, from form to transformation, to information. The system's "matter" has changed "phase", at least since Bergson. It's more liquid than solid, more airlike than liquid, more informational than material. The global is fleeing towards the fragile, the weightless, the living, the breathing' (Serres and Latour 1992, 121). There is also the recurrent theme of hard versus soft in Serres's work (e.g. Serres 1985).

Chapter 2

1 Canguilhem and Monod did in fact meet later on, and in contrast to how Serres often portrayed Canguilhem's philosophy, the latter did engage with the new developments in molecular biology from the 1960s on (see Talcott 2014; Erdur 2018).
2 This formulation is nowhere found in the work of Bachelard in this literal sense. Closer to this formulation are certain remarks by Eduard Le Roy (see Le Roy 1899).
3 Moreover, these essays might also be targeted because they discuss the case of Louis Pasteur, one of Latour's favourite authors. Canguilhem is criticized numerous times in Latour's book on Pasteur (Latour 1984, 31, 75).

Chapter 3

1 I translate *esprit* here with spirit, rather than mind. Contrary to 'mind', 'spirit' suggests that it not a matter of empirical psychology. We will see that, for Bachelard, the scientific spirit always already refers to a *transformed* psychology.
2 This is a reference to Leibniz, who similarly recounts Nolant de Fatouiville's Commedia dell'Arte play *Arlequin, Empereur de la lune* (1693). It is also the opening quote of Serres's doctoral dissertation on Leibniz.

Chapter 4

1 There is also the possibility that information might lead to an increase in possible answers, if the added information consists rather in additional possible answers. As we will see, this occurs in scientific disputes when opponents of a certain theory try to introduce alternative possible explanations of certain phenomena.
2 Dagognet writes: 'Nature, for its part, seems to work in the manner of a typographer. With limited characters, she composes a multi-color text and draws various drawings' (Dagognet 1973, 147). Derrida, similarly, argues that writing can be generalized 'to describe not only the system of notation secondarily connected with these activities but the essence and the content of these activities themselves. It is also in this sense that the contemporary biologist speaks of writing and pro-*gram* in relation to the most elementary processes of information within the living cell' (Derrida 1967, 9).

Chapter 5

1 As Schmidgen notes, they would meet once again for Lyotard's exhibition *Les immatériaux* at the Centre George Pompidou, where 'Latour participated in a computer-assisted collaborative writing project along with Michel Butor, Jacques Derrida, Isabelle Stengers, and others, which sought to explore possible collective definitions of philosophical concepts' (Schmidgen 2014, 150).
2 This in an expression that Latour and Serres use, inspired by the work of Charles Péguy (1961). Latour has been strongly inspired by Péguy, and even published on his work (Latour 2015b; see also Schmidgen 2013).

Chapter 6

1 Dumézil was similarly elected to the *Académie Française*, but in 1978, with Lévi-Strauss delivering the reception speech. Dumézil passed away in 1986 before Serres

or Girard were elected (in 1990 and 2005). Girard never extensively discussed Dumézil's work (though see Girard 1972, 41; 1982, 66–67).
2   Bruno Latour similarly argues that what distinguishes baboons from humans is the role of quasi-objects, who pacify and stabilize social relations (see Strum and Latour 1987).

Chapter 7

1   It should therefore not be confused with the ambition to purify science from religious themes, as was for instance the project of Thomas Henry Huxley and others, who aimed to show how Darwinism debunked all ideas from natural theology. Their aim was to obtain a 'science untainted by religion' (Harrison 2017, 53).
2   One of the central claims of Gauchet is that Christianity plays a crucial role in the formation of Western democratic societies (see Cloots 2015). There is a similarity with the theory of Girard, though they hardly ever refer to one another (see Tarot 2008, 624-625). An exception is an interview, in which Gauchet dissociates himself from Girard: 'There are some overlapping elements and a lot of interesting analyzes in Girard. But his vision of the universal mechanism and mimetic desire leaves me perplexed. They don't shed any light on the things I'm trying to understand and seem pretty trivial to me. This idea of our constant confrontation with a scapegoat hardly convinces me. I nevertheless agree with his diagnosis that Christianity represents a break with archaic sacredness, but I take a completely different reading of it in detail' (Giroux 2014).

Chapter 8

1   Nevertheless, there are also some clear criticisms of the term (e.g., Baskin 2015; Bonneuil and Fressoz 2016).
2   Again this comes close to a theme picked up in Isabelle Stengers's work, namely what she calls an *ecology of practices* (Stengers 2013; see Simons 2019).
3   This claim is in fact too strong, as I have argued elsewhere (see Simons 2016). Latour mixes up two possible conclusions: stating that we are *certain that we are uncertain*, which does not follow, the real conclusion is that we are *uncertain that we are certain*. We can never be certain again that we can reduce particular quasi-objects to objects, although it might work for some cases.

Bibliography

Akrich, M., M. Callon, and B. Latour. *Sociologie de la traduction: Textes fondateurs*. Paris: Presses de l'Ecole des Mines, 2006.
Althusser, L. *Pour Marx*. Paris: Maspero, 1965.
Althusser, L. *Philosophy and the Spontaneous Philosophy of the Scientists and Other Essays*. London/New York: Verso, 1974a [1990].
Althusser, L. *Essays in Self-Criticism*. London: New Left, 1974b [1976].
Althusser, L. *Sur la reproduction*. Paris: PUF, 1995.
Althusser, L., and E. Balibar. *Lire le Capital, II*. Paris: Maspero, 1965.
Assad, M. *Reading with Michel Serres: An Encounter with Time*. Albany: State University of New York Press, 1999.
Assmann, J. *Moses the Egyptian: The Memory of Egypt in Western Monotheism*. Cambridge: Harvard University Press, 1998.
Atlan, H. *Entre le cristal et la fumée: Essai sur l'organisation du vivant*. Paris: Seuil, 1979.
Atlan, H. 'Founding Violence and Divine Referent'. In *Violence and Truth: On the Work of René Girard*, edited by P. Dumouchel, 192–208. London: Athlone, 1988.
Attali, J. *La parole et l'outil*. Paris: PUF, 1975.
Bachelard, G. *Etude sur l'évolution d'un problème de physique: La propagation thermique dans les solides*. Paris: Vrin, 1927.
Bachelard, G. *Le nouvel esprit scientifique*. Paris: Alcan, 1934.
Bachelard, G. *La dialectique de la durée*. Paris: Boivin, 1936.
Bachelard, G. *La formation de l'esprit scientifique: Contribution à une psychanalyse de la connaissance objective*. Paris: Vrin, 1938a.
Bachelard, G. *The Psychoanalysis of Fire*. London: Quartet Books, 1938b [1987].
Bachelard, G. *La philosophie du non: Essai d'une philosophie du nouvel esprit scientifique*. Paris: PUF, 1940.
Bachelard, G. *L'eau et les rêves: Essai sur l'imagination de la matière*. Paris: Corti, 1942.
Bachelard, G. *L'air et les songes: Essai sur l'imagination du mouvement*. Paris: Corti, 1943.
Bachelard, G. *Le rationalisme appliqué*. Paris: PUF, 1949.
Bachelard, G. *L'activité rationaliste de la physique contemporaine*. Paris: PUF, 1951.
Bachelard, G. *Le matérialisme rationnel*. Paris: PUF, 1953.
Bachelard, G. *La poétique de l'espace*. Paris: PUF, 1957.
Bachelard, G. *La poétique de la rêverie*. Paris: PUF, 1960.
Bachelard, G. *Etudes*. Paris: Vrin, 1970.
Bachelard, G. *L'engagement rationaliste*. Paris: PUF, 1972.
Baird, D. *Thing Knowledge: A Philosophy of Scientific Instruments*. Berkeley: University of California Press, 2004.

Balibar, E. 'From Bachelard to Althusser: The Concept of "Epistemological Break"'. *Economy and Society* 7, no. 3 (1978): 207–37.

Barnes, B., and D. Edge. *Science in Context: Readings in the Sociology of Science*. Milton Keynes: Open University Press, 1982.

Barnes, B., D. Bloor, and J. Henry. *Scientific Knowledge: A Sociological Analysis*. London: Athlone,1996.

Baskin, J. 'Paradigm Dressed as Epoch: The Ideology of the Anthropocene'. *Environmental Values* 24, no. 1 (2015): 9–29.

Beaune, J. 'François Dagognet: de la matière a l'objet'. In *François Dagognet: Un nouvel Encyclopédiste?*, edited by D. Parrochia, 25–57. Seyssel: Champ Vallon, 2011.

Belier, W. *Decayed Gods: Origin and Development of Georges Dumezil's 'Idéologie tripartie'*. Leiden: Brill, 1991.

Bensaude-Vincent, B. 'Michel Serres, Historien Des Sciences'. In *Le Cahier de l'Herne: Michel Serres*, edited by F. Yvonnet, 37–46. Paris: Editions de l'Herne, 2010.

Berger, P. *The Sacred Canopy: Elements of a Sociological Theory of Religion*. Garden City: Doubleday, 1967.

Bloor, D. *Knowledge and Social Imagery*. London: Routledge and Kegan Paul, 1976.

Bonneuil, C., and J. Fressoz. *The Shock of the Anthropocene: The Earth, History and Us*. London: Verso, 2016.

Bordoni, S. *When Historiography Met Epistemology*. Leiden: Brill, 2017.

Bourdieu, P. 'The Specificity of the Scientific Field and the Social Conditions of the Progress of Reason'. *Social Science Information* 14, no. 6 (1975): 19–47.

Bourdieu, P. *Science of Science and Reflexivity*. Cambridge: Polity Press, 2004.

Bowker, G., and B. Latour. 'A Booming Discipline Short of Discipline: (Social) Studies of Science in France'. *Social Studies of Science* 17, no. 4 (1987): 715–48.

Brenner, A. *Les origines françaises de la philosophie des sciences*. Paris: PUF, 2003.

Brenner, A. 'Is There a Cultural Barrier Between Historical Epistemology and Analytic Philosophy of Science?' *International Studies in the Philosophy of Science* 29, no. 2 (2015): 201–14.

Brillouin, L. *Science and Information Theory*. New York: Academic Press, 1956.

Brown, S. 'Michel Serres: Science, Translation and the Logic of the Parasite'. *Theory, Culture and Society* 19, no. 3 (2002): 1–27.

Brunschvicg, L. *De la connaissance de soi*. Paris: Alcan, 1931.

Bühlmann, V. *Mathematics and Information in the Philosophy of Michel Serres*. London: Bloomsbury, 2020.

Callon, M. 'Sociologie des techniques?' *Pandore* 2 (1975): 28–32.

Callon, M. 'Struggles and Negotiations to Define What Is Problematic and What Is Not: The Socio-logic of Translation'. In *The Social Process of Scientific Investigation, Sociology of the Sciences, a Yearbook*, edited by K. Knorr, R. Krohn and R. Whitley, 197–219. Dordrecht: Reidel, 1981a.

Callon, M. 'Boites Noires et Opérations de Traduction'. *Economie et Humanisme* 262 (1981b): 53–9.

Callon, M. 'Some Elements of a Sociology of Translation: Domestication of the Scallops and the Fishermen of St Brieuc Bay'. *The Sociological Review* 32 (1984): 196–233.

Callon, M. 'The Sociology of an Actor-Network: The Case of the Electric Vehicle'. In *Mapping the Dynamics of Science and Technology*, edited by M. Callon., J. Law, and A. Rip, 19–34. London: Macmillan, 1986.

Callon, M., and B. Latour. 'Unscrewing the Big Leviathan: How Actors Macro-Structure Reality and How Sociologists Help Them to Do So'. In *Advances in Social Theory and Methodology. Toward an Integration of Micro- and Macro-Sociologies*, edited by K. Knorr-Cetina and A. Cicourel, 277–303. London: Routledge, 1981.

Callon, M., J. Law and A. Rip, edited by *Mapping the Dynamics of Science and Technology*. London: Macmillan, 1986.

Canguilhem, G. *La formation du concept de réflexe aux XVIIe et XVIIIe siècles*. Paris: PUF, 1955.

Canguilhem, G. *Ideology and Rationality in the History of the Life Sciences*. Cambridge: MIT Press, 1977 [1988].

Canguilhem, G. *Vie et mort de Jean Cavaillès*. Ambialet: Pierre Laleure, 1984.

Canguilhem, G. *The Normal and the Pathological*. New York: Zone Books, 1989.

Carroll, S. *Brave Genius: A Scientist, a Philosopher, and Their Daring Adventures from the French Resistance to the Nobel Prize*. New York: Crown Publishers, 2013.

Cassou-Noguès, P., and P. Gillot. *Le concept, le sujet et la science: Cavaillès, Canguilhem, Foucault*. Paris: Vrin, 2009.

Castelli Gattinara, E. *Les inquiétudes de la raison: épistémologie et histoire en France dans l'entre-deux-guerres*. Paris: Vrin, 1998.

Cavaillès, J. *Remarques sur la formation de la théorie abstraite des ensembles*. Paris: Hermann, 1938.

Cavaillès, J. *Sur la logique et la théorie de la science*. Paris: PUF, 1947.

Chamak, B. *Le Groupe Des Dix, ou les avatars des rapports entre science et politique*. Paris: Éditions du Rocher, 1997.

Chartier, E. 'Commentaire aux fragments de Jules Lagneau'. *Revue de métaphysique et de morale* 6, no. 5 (1898): 529–65.

Chateauraynaud, F. 'Forces et faiblesses de la nouvelle anthropologie des sciences'. *Critique* 529–530, no. 47 (1991): 459–78.

Chimisso, C. *Gaston Bachelard: Critic of Science and the Imagination*. London: Routledge, 2001.

Chimisso, C. *Writing the History of the Mind: Philosophy and Science in France, 1900 to 1960s*. Aldershot: Ashgate, 2008a.

Chimisso, C. 'From Phenomenology to Phenomenotechnique: The Role of Early Twentieth-Century Physics in Gaston Bachelard's Philosophy'. *Studies in History and Philosophy of Science*, 39, no. 3 (2008b): 384–92.

Chimisso, C. 'Narrative and Epistemology: Georges Canguilhem's Concept of Scientific Ideology'. *Studies in History and Philosophy of Science* 54 (2015): 64–73.

Cloots, A. 'Christianity, Incarnation and Disenchantment: Marcel Gauchet on the "Departure From Religion"'. In *Radical Secularization? An Inquiry into the Religious Roots of Secular Culture*, edited by S. Latré, W. Van Herck, and G. Vanheeswijck, 47–66. New York: Bloomsbury, 2015.

Collins, H., and S. Yearley. 'Epistemological Chicken'. In *Science as Practice and Culture*, edited by A. Pickering, 301–26. Chicago: University of Chicago Press, 1992.

Crahay, A. *Michel Serres. La mutation du cogito: Genèse du transcendantal objectif.* Bruxelles: De Boeck, 1988.

Cremonesi, L., O. Irrera, D. Lorenzini, and M. Tazzioli. *Foucault and the Making of Subjects*. Lanham: Rowman and Littlefield, 2016.

Crutzen, P. 'Geology of Mankind'. *Nature* 415, no. 6867 (2002): 23.

Dagognet, F. *La raison et les remèdes*. Paris: PUF, 1964.

Dagognet, F. *Méthodes et doctrine dans l'œuvre de Pasteur*. Paris: PUF, 1967.

Dagognet, F. *Tableaux et langages de la chimie*. Paris: Seuil, 1969.

Dagognet, F. *Écriture et iconographie*. Paris: Vrin, 1973.

Dagognet, F. *Pour une théorie générale des formes*. Paris: Vrin, 1975.

Dagognet, F. *Mémoire pour l'avenir: Vers une méthodologie de l'informatique*. Paris: Vrin, 1979.

Dagognet, F. *Rematérialiser: Matières et matérialismes*. Paris: Vrin, 1985.

Dagognet, F. *Eloge de l'objet: Pour une philosophie de la marchandise*. Paris: Vrin, 1989.

Dagognet, F. 'Ouverture'. In *François Dagognet, médecin épistémologue philosophe: Une philosophie à l'œuvre*, edited by R. Damien, 15–25. Le Plessis-Robinson: Institut Synthélabo pour le progrès de la connaissance, 1998.

Daston, L. 'Taking Note(s)'. *Isis* 95, no. 3 (2004): 443–8.

Daston, L., and P. Galison. *Objectivity*. Brooklyn: Zone Books, 2007.

De Rosnay, J. *Le macroscope: Vers une vision globale*. Paris: Seuil, 1975.

Deleuze, G., and F. Guattari. *Anti-Oedipus: Capitalism and Schizophrenia*. London: Continuum, 1972 [2004].

Deleuze, G., and F. Guattari. *A Thousand Plateaus*. New York: Continuum, 1980 [2004].

Delco, Alessandro. *Morphologies : à Partir Du Premier Serres*. Paris: Kimé, 1998.

Derrida, J. *Of Grammatology*. Baltimore: Johns Hopkins University Press, 1967 [1976].

Despret, V. 'The Engima of the Raven'. *Angelaki* 20, no. 2 (2015): 57–72.

Dicks, H. 'Physics, Philosophy and Poetics at the End of the Groupe des Dix: Edgar Morin and Michel Serres on the Nature of Nature'. *Natures Sciences Sociétés* 27, no. 2 (2019): 169–77.

Dolphijn, R. 'Introduction: Michel Serres and the Times'. In *Michel Serres and the Crises of the Contemporary*, edited by Rick Dolphijn, 1–9. London: Bloomsbury, 2018.

Duby, G. *The Three Orders: Feudal society Imagined*. Chicago: University of Chicago Press, 1978 [1981].

Dumézil, G. 'La préhistoire des flamines majeurs'. *Revue De L'histoire Des Religions* 118 (1938): 188–200.

Dumézil, G. *Jupiter, Mars, Quirinus. Essai sur la conception indo-européenne de la société et sur les origines de Rome*. Paris: Gallimard, 1941.

Dumezil, G. 'Les trois fonctions dans quelques traditions grecques'. In *Hommages à Lucien Febvre, II*, 25–32. Paris: Colin,1953.

Dumézil, G. *L'idéologie tripartie des Indo-Européens*. Bruxelles: Collection Latomus, XXXI, 1958.

Dumézil, G. *Mythe et épopée I. L'idéologie des trois fonctions dans les épopées des peules indo-européens*. Paris: Gallimard, 1968.

Dumézil, G. *Entretiens avec Didier Eribon*. Paris: Gallimard, 1987.

Dumouchel, P., and J. Dupuy. *L'enfer des choses: René Girard et la logique de l'économie*. Paris: Seuil, 1979.

Dupuy, J. *Ordres et désordres: Enquête sur un nouveau paradigme*. Paris: Seuil, 1982a.

Dupuy, J. 'Mimésis et morphogénèse'. In *René Girard et le problème du mal*, edited by M. Deguy and J. Dupuy, 225–78. Paris: Grasset, 1982b.

Dupuy, J. 'Totalization and Misrecognition'. In *Violence and Truth: On the Work of René Girard*, edited by P. Dumouchel, 75–100. London: Athlone, 1988.

Dupuy, J. *Pour un catastrophisme éclairé*. Paris: Seuil, 2002.

Dupuy, J. 'Naturalizing Mimetic Theory'. In *Mimesis and Science*, edited by S. Garrels, 193–213. East Lansing: Michigan State University Press, 2011.

Dupuy, J. *The Mark of the Sacred*. Stanford: Stanford University Press,2013.

Dupuy, J., and S. Karsenty. *L'Invasion pharmaceutique*. Paris: Seuil, 1977.

Dupuy, J., and J. Robert. *La trahison de l'opulence*. Paris: PUF, 1976.

Elden, S. *Foucault's Last Decade*. Chichester: Wiley-Blackwell, 2016.

Erdur, O. *Die epistemologischen Jahre. Philosophie und Biologie in Frankreich, 1960–1980*. Zürich: Chronos Verlag, 2018.

Eribon, D. *Faut-il brûler Dumézil? Mythologie, science et politique*. Paris: Flammarion, 1992.

Eribon, D. *Michel Foucault et ses contemporains*. Paris: Fayard, 1994.

Ferraris, M. *Manifesto of New Realism*. New York: State University of New York Press, 2014.

Ferry, L. and M. Gauchet. *Le religieux après la religion*. Paris: Grasset, 2004.

Foucault, M. 'Madness Only Exists in Society'. In *Foucault Live: Collected Interviews, 1961–1984*, trans. Lysa Hochroth, 7–9. New York: Semiotext(e), 1961 [1996].

Foucault, M. 'Truth and Juridical Forms'. In *Power. Volume 3 of Essential Works of Foucault: 1954–1984*, edited by James D. Faubion, 1–89. New York: The New Press, 1973 [2000].

Foucault, M. *Surveiller et punir: Naissance de la prison*. Paris: Gallimard, 1975.

Foucault, M. *The Foucault Reader*. New York: Pantheon Books, 1984a.

Foucault, M. *Histoire de la sexualité. 2: L'usage des plaisirs*. Paris: Gallimard, 1984b.

Foucault, M. *Histoire de la sexualité. 3: Le souci de soi*. Paris: Gallimard, 1984c.

Foucault, M. *Technologies of the Self: A Seminar with Michel Foucault*. Amherst: University of Massachusetts, 1988.

Foucault, M. 'Introduction'. In *The Normal and the Pathological*, edited by G. Canguilhem, xi–xx. New York: Zone Books, 1989.
Foucault, M. *The Hermeneutics of the Subject*. Basingstoke: Palgrave, 2005.
Foucault, M. *The Government of Self and Others*. Basingstoke: Palgrave Macmillan, 2010.
Foucault, M. *About the Beginning of the Hermeneutics of the Self: Lectures at Darthmouth College, 1980*. Chicago: University of Chicago Press, 2015.
Foucault, M. *Histoire de la sexualité. 4: Les Aveux de la chair*. Paris: Gallimard, 2018.
Frémont, C. 'Bachelard et Michel Serres: deux tiers-instruits?' *Cahiers Gaston Bachelard* 10 (2008): 75–88.
Frémont, C. 'Philosophie pour le temps présent'. In *Le Cahier de l'Herne: Michel Serres*, edited by F. Yvonnet, 17–26. Paris: Editions de l'Herne, 2010.
Fuller, S. 'The Science Wars: Who Exactly is the Enemy?' *Social Epistemology* 13, no. 3–4 (1999): 243–9.
Fuller, S. *The Governance of Science: Ideology and The Future of the Open Society*. Buckingham: Open University Press, 2000.
Fuller, S. *Social Epistemology*. 2nd edn. Bloomington: Indiana University Press, 2002.
Galison, P. 'Image of the Self'. In *Things That Talk: Object lessons from Art and Science*, edited by L. Daston, 257–94. New York: Zone Books, 2004.
Gane, M. *Auguste Comte*. London: Routledge, 2006.
Gauchet, M. *The Disenchantment of the World: A Political History of Religion*. Princeton: Princeton University Press, 1985 [1999].
Gauchet, M. *La religion dans la démocratie: Parcours de la laïcité*. Paris: Gallimard, 1998.
Gauchet, M. *La condition historique*. Paris: Plon, 2003.
Geroulanos, S., and J. Phillips. 'Eurasianism versus Indo-Germanism: Linguistics and Mythology in the 1930s' Controversies over European Prehistory'. *History of Science* 56, no. 3 (2018): 343–78.
Gingras, Y. 'Following Scientists through Society? Yes, but at Arm's Length!'. In *Scientific Practice: Theories and Stories of Doing Physics*, edited by J. Buchwald, 123–48. Chicago: University of Chicago Press, 1995.
Girard, R. *Violence and the Sacred*. London: Bloomsbury, 1972 [2017].
Girard, R. *The Scapegoat*. London: Athlone, 1982 [1986].
Girard, R. 'Introduction'. In *Detachment*, edited by Michel Serres, vii–ix. Athens: Ohio University Press, 1989.
Girard, R. *I See Satan Fall Like Lightning*. Maryknoll, MD: Orbis, 2001.
Girard, R. 'From Ritual to Science'. In *Mapping Michel Serres*, edited by N. Abbas, 10–23. Ann Arbor: University of Michigan Press, 2005.
Girard, R. *Battling to the End: Conversations with Benoît Chantre*. East Lansing: Michigan State University Press, 2007 [2010].
Girard, R. *Evolution and Conversion: Dialogues on the Origins of Culture*. London: Continuum, 2008.

Girard, R. 'Mimesis and Science: An Interview with René Girard'. In *Mimesis and Science*, edited by S. Garrels, 215–53. East Lansing: Michigan State University Press, 2011.

Girard, R., and M. Serres. *Le tragique et la pitié: Discours de réception de René Girard à l'Académie française et réponse de Michel Serres*. Paris: Le Pommier, 2007.

Giroux, M. 'Entretien avec Marcel Gauchet: « Le monde moderne est sous le signe de l'ignorance'. *Phillit*, 2014. Consulted at 10 May 2021. http://phillitt.fr/2014/10/09/entretien-avec-marcel-gauchet-le-monde-moderne-est-sous-le-signe-de-lignorance/

Godin, C. 'Panorama d'une pensée'. In *Le Cahier de l'Herne: Michel Serres*, edited by F. Yvonnet, 27–34. Paris: Editions de l'Herne, 2010.

Godin, C. 'François Dagognet: Tendances encyclopédiques et tentations du système'. In *François Dagognet: Un nouvel encyclopédiste?*, edited by D. Parrochia, 77–84. Seyssel: Champ Vallon, 2011.

Goldstein, J. *The Post-Revolutionary Self: Politics and Psyche in France, 1750–1850*. Cambridge: Harvard University Press, 2005.

Golinski, J. *Making Natural Knowledge: Constructivism and The History of Science*. Cambridge: Cambridge University Press, 1998.

Granger, G. 'Jean Cavaillès ou la montée vers Spinoza'. *Les études Philosophiques* 2, no. 3/4 (1947): 271–9.

Gratton, P. *Speculative Realism: Problems and Prospects*. London: Bloomsbury, 2014.

Greimas, A. *Sémiotique et sciences sociales*. Paris: Le Seuil, 1976.

Hacking, I. *The Social Construction of What?*. Cambridge: Harvard University Press, 1999.

Hadot, P. *What Is Ancient Philosophy?* Cambridge: Harvard University Press, 2002.

Harari, J. and D. Bell. 'Introduction: Journal à plusieurs voies'. In *Hermes: Literature, Science, Philosophy*, edited by Josué Harari and David Bell, ix–xl. Baltimore: The Johns Hopkins University Press, 1982.

Haraway, D. 'Situated Knowledges: The Science Question in Feminism and the Privilege of Partial Perspective'. *Feminist Studies* 14, no. 3 (1988): 575–99.

Haraway, D. *Simians, Cyborgs, and Women: The Reinvention of Nature*. London: Free Association, 1991.

Harding, S. *The Science Question in Feminism*. Milton Keynes: Open University Press, 1986.

Harman, G. *Prince of Networks: Bruno Latour and Metaphysics*. Melbourne: Re.press, 2009.

Harman, G. *Object-Oriented Ontology: A New Theory of Everything*. London: Pelican, 2018a.

Harman, G. *Speculative Realism: An Introduction*. Cambridge: Polity Press, 2018b.

Harris, P. 'The Itinerant Theorist: Nature and Knowledge/Ecology and Topology in Michel Serres'. *SubStance* 26, no. 2 (1997): 37–58.

Harrison, P. 'Science and Secularization'. *Intellectual History Review* 27, no. 1 (2017): 47–70.

Haven, C. *Evolution of Desire: A Life of René Girard*. East Lansing: Michigan State University Press, 2018.
Hepler-Smith, E. 'Paper Chemistry: François Dagognet and the Chemical Graph'. *Ambix* 65, no. 1 (2018): 76–98.
Herzogenrath, B., edited by *Time and History in Deleuze and Serres*. London: Continuum, 2012.
Illich, I. *Deschooling Society*. Harmondsworth: Penguin Books, 1971.
Illich, I. *Energy and Equity*. New York: Harper and Row Publishers, 1974.
Illich, I. *Medical Nemesis: The Expropriation of Health*. London: Calder and Boyars, 1975.
Illich, I. *Shadow Work*. London: Marion Boyars, 1981.
Jacob, F. *La logique du vivant: Une histoire de l'hérédité*. Paris: Gallimard, 1970.
Jacob, F. *Le jeu des possibles: Essai sur la diversité du vivant*. Paris: Fayard, 1981.
Jacob, F. *La statue intérieure*. Paris: Seuil, 1987.
Jones, M. *The Good Life in the Scientific Revolution: Descartes, Pascal, Leibniz and the Cultivation of Virtue*. Chicago: University of Chicago Press, 2006.
Karaca, K. 'A Case Study in Experimental Exploration: Exploratory Data Selection at the Large Hadron Collider'. *Synthese* 194, no. 2 (2017): 333–54.
Keeling, C. 'Rewards and Penalties of Monitoring the Earth'. *Annual Review of Energy and the Environment* 23, no. 1 (1998): 25–82.
Klein, U. *Experiments, Models, Paper Tools: Cultures of Organic Chemistry in the Nineteenth Century*. Stanford: Stanford University Press, 2003.
Knorr, K. 'Producing and Reproducing Knowledge: Descriptive or Constructive? Towards a Model of Research Production'. *Social Science Information* 16, no. 6 (1977): 669–96.
Koenig, M. 'Beyond the Paradigm of Secularization?'. In *Working with a Secular Age: Interdisciplinary Perspectives on Charles Taylor's Master Narrative*, edited by F. Zemmin, C. Jager, and G. Vanheeswijck, 23–48. Berlin: De Gruyter, 2016.
Kojève, Alexandre. 'L'origine chrétienne de la science moderne,' In *Mélanges Alexandre Koyré. 2 : L'aventure De L'esprit*, edited by Braudel Fernand, 296–306. Paris, Hermann, 1964.
Kuhn, T. *The Structure of Scientific Revolutions*. Chicago: University of Chicago Press, 1962.
Kuhn, T. *The Road since Structure: Philosophical Essays, 1970–1993*. Chicago: University of Chicago Press, 2000.
Kukla, A. *Social Constructivism and the Philosophy of Science*. London: Routledge, 2000.
Ladrière, J. 'Préface'. In *La mutation du cogito: Genèse du transcendantal objectif*, edited by A. Crahay and Michel Serres, 9–15. Bruxelles: De Boeck,1988.
Latour, B. 'Is it Possible to Reconstruct the Research Process? Sociology of a Brain Peptide'. In *The Social Process of Scientific Investigation, Sociology of the Sciences, a Yearbook*, edited by K. Knorr, R. Krohn and R. Whitley, 53–77. Dordrecht: Reidel, 1981.

Latour, B. 'Give Me a Laboratory and I will Raise the World'. In *Science Observed: Perspectives on the Social Study of Science*, edited by K. Knorr-Cetina and M. Mulkay, 141-70. London: Sage, 1983.

Latour, B. *The Pasteurization of France*. London: Harvard University Press, 1984 [1988].

Latour, B. 'Visualization and Cognition'. *Sociology of Knowledge. Studies in the Sociology of Culture* 6 (1986): 1-40.

Latour, B. *Science in Action: How to Follow Scientists and Engineers Through Society*. Cambridge: Harvard University Press,1987a.

Latour, B. 'The Enlightenment without the Critique: A Word on Michel Serres' Philosophy'. *Royal Institute of Philosophy Lecture Series* 21 (1987b): 83-97.

Latour, B. 'Postmodern? No, Simply Amodern! Steps towards an Anthropology of Science'. *Studies in History and Philosophy of Science* 21, no. 1 (1990a): 145-71.

Latour, B. 'Force and Reason of Experiment'. In *Experimental Inquiries: Historical, Philosophical and Social Studies of Experimentation in Science*, edited by H. Le Grand, 49-80. Dordrecht: Kluwer, 1990b.

Latour, B. *We Have Never Been Modern*. London: Prentice Hall, 1991 [1993].

Latour, B. (1998). From the World of Science to the World of Research? *Science* (American Association for the Advancement of Science), 280(5361), 208-209. https://doi.org/10.1126/science.280.5361.208

Latour, B. *Politics of Nature: How to Bring the Sciences into Democracy*. Cambridge: Harvard University Press, 1999a [2004].

Latour, B. *Pandora's Hope: Essays on the Reality of Science Studies*. Cambridge: Harvard University Press, 1999b.

Latour, B. 'For David Bloor ... and beyond: A Reply to David Bloor's "Anti-Latour"'. *Studies in History and Philosophy of Science* 30, no. 1 (1999c): 113-29.

Latour, B. 'On Recalling ANT'. *Sociological Review* 47, no. S1 (1999d): 15-25.

Latour, B. 'Réponse aux objections...'. *Revue Du MAUSS* 17, no. 1 (2001): 137-52.

Latour, B. 'Il ne faut plus qu'une science soit ouverte ou fermée'. *Rue Descartes* 41 (2003): 66-81.

Latour, B. 'How to Talk About the Body? The Normative Dimension of Science Studies'. *Body and Society* 10, no. 2-3 (2004): 205-29.

Latour, B. 'From *Realpolitik* to *Dingpolitik* or How to Make Things Public in Making Things Public'. In *Making Things Public: Atmospheres of Democracy*, edited by B. Latour and P. Weibel, 14-41. Karlsruhe: ZKM Center for Art and Media, 2005a.

Latour, B. *Reassembling the Social: An Introduction to Actor-Network Theory*. Oxford: Oxford University Press, 2005b.

Latour, B. 'Entretien avec Bruno Latour'. *Tracés: Revue de Sciences humaines* 10 (2006): 113-29.

Latour, B. *Chroniques d'un amateur de sciences*. Paris: Presses de Mines,2006.

Latour, B. '5 Questions'. In *Philosophy of Technology: 5 Questions*, edited by J. Olsen, K. Berg, and E. Selinger, 125-35. New York: Automa,2007.

Latour, B. *Cogitamus: Six lettres sur les humanités scientifiques*. Paris: La Découverte, 2010a.

Latour, B. *On the Modern Cult of the Factish Gods*. Durham, NC: Duke University Press, 1996 [2010].

Latour, B. 'An Attempt at a Compositionist Manifesto'. *New Literary History* 41, no. 3 (2010b): 471–90.

Latour, B. *An Inquiry into Modes of Existence: An Anthropology of the Moderns*. Cambridge: Harvard University Press, 2013a.

Latour, B. 'Biography of an Inquiry: On a Book about Modes of Existence'. *Social Studies of Science* 43, no. 2 (2013b): 287–301.

Latour, B. 'Agency at the Time of the Anthropocene'. *New Literary History: A Journal of Theory and Interpretation* 45, no. 1 (2014): 1–18.

Latour, B. *Facing Gaia: Eight Lectures on the New Climatic Regime*. Cambridge: Polity 2015a [2017].

Latour, B. 'Charles Péguy: Time, Space, and le Monde Moderne'. *New Literary History* 46, no. 1 (2015b): 41–62.

Latour, B., and T. Crawford. 'An Interview with Bruno Latour'. *Configurations* 1, no. 2 (1993): 247–68.

Latour, B., and P. Fabbri. 'Rhétorique de la science'. *Actes de la recherche en sciences sociales* 13 (1977): 81–95.

Latour, B., and S. Woolgar. *Laboratory Life: The Social Construction of Scientific Facts*. Beverly Hills: Sage, 1979.

Le Roy, E. 'Science et philosophie (Suite)'. *Revue de métaphysique et de morale* 7, no. 5 (1899): 503–62.

Lecourt, D. *L'épistémologie historique de Gaston Bachelard*. Paris: Vrin, 1969.

Lecourt, D. *Bachelard ou Le jour et la nuit: Un essai du matérialisme dialectique*. Paris: Grasset, 1974.

Lecourt, D. *Marxism and epistemology: Bachelard, Canguilhem and Foucault*. London: NLB, 1975.

Lecourt, D. *Lyssenko: Histoire réelle d'une science prolétarienne*. Paris: Maspero, 1976.

Lehtonen, T. 'Serres and Foundations'. *Theory, Culture and Society* 37, no. 3 (2020): 3–22.

Loeve, S., X. Guchet, and B. Bensaude-Vincent, edited by *French Philosophy of Technology*. Cham: Springer, 2018.

Loison, L. 'Forms of Presentism in the History of Science. Rethinking the Project of Historical Epistemology'. *Studies in History and Philosophy of Science. Part A* 60 (2016): 29–37.

Lorenzini, D. *Éthique et politique de soi: Foucault, Hadot, Cavell et les techniques de l'ordinaire*. Paris: Vrin, 2015.

Luckmann, T. *The Invisible Religion: The Problem of Religion in Modern Society*. New York: Macmillan, 1967.

Lyotard, J. *Discourse, Figure*. Minneapolis: University of Minnesota Press, 1971 [2011].

Lyotard, J. *Libidinal Economy*. Bloomington: Indiana University Press, 1974 [1993].

Lyotard, J. *Instructions païennes*. Paris: Cahiers Galilée,1977a.

Lyotard, J. *Rudiments païens: Genre dissertatif*. Paris: Union générale d'éditions, 1977b.

Lyotard, J. *The Postmodern Condition: A Report on Knowledge*. Minneapolis: University of Minnesota, 1979 [1984].

Lyotard, J. *The Differend: Phrases in Dispute*. Manchester: Manchester University Press, 1983 [1988].

Lyotard, J. *The Postmodern Explained: Correspondence 1982-1985*. Minneapolis: University of Minnesota, 1986 [1993].

Lyotard, J. 'Dialogue pour un temps de crise (Interview Collective)'. *Le Monde*, 15 April 1988, p. Xxxviii,1988a.

Lyotard, J. *The Inhuman: Reflections on Time*. Cambridge: Polity Press, 1988b [1991].

Lyotard, J. *Peregrinations: Law, Form, Event*. New York: Columbia University Press, 1988c.

Lyotard, J. 'The Sign of History'. In *The Lyotard Reader*, edited by A. Benjamin, 393-411. Oxford: Blackwell, 1989.

Lyotard, J. 'Humor in Semiotheology'. In *Toward the Postmodern*, edited by R. Harvey and M. Roberts, 73-86. New Jersey: Humanities Press, 1993.

Lyotard, J., and J. Thébaud. *Just Gaming*. Minneapolis: University of Minnesota, 1979 [1985].

Macherey, P. 'Althusser and the Concept of the Spontaneous Philosophy of Scientists'. *Parrhesia* 6 (2009): 14-27.

Marks, J. 'Jacques Monod, François Jacob, and the Lysenko Affair'. *L'Esprit Créateur* 52, no. 2 (2012): 75-88.

Mazanderan, F. and B. Latour. 'The Whole World is Becoming Science Studies: Fadhila Mazanderani Talks with Bruno Latour'. *Engaging Science, Technology, and Society* 4 (2018): 284-302.

McGushin, E. *Foucault's Askesis: An Introduction to the Philosophical Life*. Evanston: Northwestern University Press, 2007.

Merton, R. 'The Self-Fulfilling Prophecy'. *The Antioch Review* 8, no. 2 (1948): 193-210.

Méthot, P. 'On the Genealogy of Concepts and Experimental Practices: Rethinking Georges Canguilhem's Historical Epistemology'. *Studies in History and Philosophy of Science* 44, no. 1 (2013): 112-23.

Mialet, H. 'Where would STS be without Latour? What would be Missing?' *Social Studies of Science* 42, no. 3 (2012): 456-61.

Mikami, K. and S. Woolgar. 'STS as a Program of Ontological Disobedience: Koichi Mikami Talks with Steve Woolgar'. *Engaging Science, Technology, and Society* 4 (2018): 303-19.

Momigliano, A. 'Georges Dumézil and the Trifunctional Approach to Roman Civilization'. *History and Theory* 23, no. 3 (1984): 312-30.

Monod, J. *Le hasard et la nécessité: Essai sur la philosophie naturelle de la biologie moderne*. Paris: Seuil, 1970.

Morin, E. *La méthode: 1: La nature de la nature*. Paris: Seuil, 1977.

Morin, E., R. Passet, F. Vivien, and H. Dicks. 'Edgar Morin et René Passet: Les passeurs du Groupe des Dix'. *Natures Sciences Sociétés* 27, no. 2 (2019): 225–37.

Morton, T. 'How I Learned to Stop Worrying and Love the Term Anthropocene'. *Cambridge Journal of Postcolonial Literary Inquiry* 1, no. 2 (2014): 257–64.

Moser, K. *The Encyclopedic Philosophy of Michel Serres*. Augusta: Anaphora Literary Press, 2016.

Mouffe, C. *On the Political*. London: Routledge, 2005.

Passet, R. *L'Économique et le vivant*. Paris: Payot, 1979.

Paulson, W. 'Michel Serres's Utopia of Language'. *Configurations* 8, no. 2 (2000): 215–28.

Paulson, W. 'Swimming the Channel'. In *Mapping Michel Serres*, edited by N. Abbas, 24–36. Ann Arbor: University of Michigan Press, 2005.

Pêcheux, M. and M. Fichant. *Sur l'histoire des sciences*. Paris: François Maspero, 1969.

Peden, K. *Spinoza Contra Phenomenology: French Rationalism from Cavaillès to Deleuze*. Stanford: Stanford University Press, 2014.

Péguy, C. 'Clio. Dialogue de l'histoire et de l'arne paienne'. In *Œuvres en prose*, edited by Marcel Péguy, 93–309. Paris: Gallimard, 1961.

Pickering, M. 'Auguste Comte and the Return to Primitivism'. *Revue Internationale De Philosophie* 52, no. 1 (1998): 51–77.

Pinch, T. *Confronting Nature: The Sociology of Solar-neutrino Detection*. Dordrecht: Reidel, 1986.

Popper, K. *Conjectures and Refutations: The Growth of Scientific Knowledge*. London: Routledge and Kegan Paul, 1963.

Popper, K. 'Epistemology without a Knowing Subject'. *Studies in Logic and the Foundations of Mathematics* 52 (1968): 333–73.

Prigogine, I, and I. Stengers. *La nouvelle alliance: Métamorphose de la science*. Paris: Gallimard, 1979.

Prigogine, I. and I. Stengers. 'Dynamics from Leibniz to Lucretius'. In *Hermes: Literature, Science, Philosophy*, edited by J. Harari and D. Bell, 136–55. Baltimore: The Johns Hopkins University Press, 1982.

Resch, R. *Althusser and the Renewal of Marxist Social Theory*. Berkeley: University of California Press, 1992.

Rheinberger, H. 'Gaston Bachelard and the Notion of "Phenomenotechnique"'. *Perspectives on Science* 13, no. 3 (2005): 313–28.

Rheinberger, H. *On Historicizing Epistemology: An Essay*. Stanford: Stanford University Press, 2010.

Rheinberger, H. *Der Kupferstecher und der Philosoph. Albert Flocon trifft Gaston Bachelard*. Berlin: Diaphanes, 2016.

Rose, N. *Inventing Our Selves. Psychology, Power and Personhood*. Cambridge: Cambridge University Press, 1998.

Rouse, J. *Engaging Science: How to Understand Its Practices Philosophically*. Ithaca: Cornell University Press, 1996.

Sartre, J. *L'être et le néant: Essai d'ontologie phénoménologique*. Paris: Gallimard, 1943.

Savoia, P. 'Towards a Historical Epistemology of the Self'. *Medicinia and Storia* 6 (2014): 37–52.

Sayes, E. 'From the Sacred to the Sacred Object: Girard, Serres, and Latour on the Ordering of the Human Collective'. *Techné: Research in Philosophy and Technology* 16, no. 2 (2012): 105–22.

Scharff, R. *Comte after Positivism*. Cambridge: Cambridge University Press, 1995.

Schlerath, B. 'Georges Dumézil und die Rekonstruktion der indogermanischen Kultur, Teil 1'. *Kratylos* 40 (1995): 1–48.

Schlerath, B. 'Georges Dumézil und die Rekonstruktion der indogermanischen Kultur, Teil 2'. *Kratylos* 41 (1996): 1–67.

Schmidgen, H. 'The Materiality of Things? Bruno Latour, Charles Péguy and the History of Science'. *History of the Human Sciences* 26, no. 1 (2013): 3–28.

Schmidgen, H. *Bruno Latour in Pieces: An Intellectual Biography*. New York: Fordham University Press, 2014.

Scotti, P. *Galileo Revisited: The Galileo Affair in Context*. San Francisco: Ignatius Press, 2017.

Serres, M. *Le système de Leibniz et ses modèles mathématiques: étoiles - schémas - points*. Paris: PUF, 1968a.

Serres, M. 'Les Messager'. *Bulletin de la Société française de Philosophie* 63 (1968b): 33–71.

Serres, M. *Hermès I, La communication*. Paris: Éditions de Minuit, 1969.

Serres, M. 'Reformation and the Seven Sins'. *Parrhesia* 31 (1970 [2019]): 33–47.

Serres, M. *Hermès II, L'interférence*. Paris: Éditions de Minuit, 1972.

Serres, M. *Hermès III, La traduction*. Paris: Éditions de Minuit,1974a.

Serres, M. 'Les Sciences'. In *Faire de l'histoire, vol. 2*, edited by J. Le Goff and P. Nora, 203–28. Paris: Gallimard, 1974b.

Serres, M. 'Introduction'. In *Cours de philosophie positive, leçons 1 à 45*, edited by Auguste Comte , 1–19. Paris: Hermann, 1975a.

Serres, M. *Esthétiques sur Carpaccio*. Paris: Hermann, 1975b.

Serres, M. 'Le philosophe et la guerre – entretien de Michel Serres avec Françoise Lévy'. *Nouvelles littéraires* 54, no. 2536 (1976): 21.

Serres, M. *Hermès IV, La distribution*. Paris: Éditions de Minuit, 1977a.

Serres, M. *The Birth of Physics*. Manchester: Clinamen Press, 1977b [2000].

Serres, M. *The Parasite*. Baltimore: The John Hopkins University Press, 1980a [2007].

Serres, M. *Hermès V: Le passage du nord-ouest*. Paris: Editions de Minuit, 1980b.

Serres, M. *Genesis*. Ann Arbor: University of Michigan Press, 1982a [1999].

Serres, M. 'L'homme est un loup pour l'homme'. In *René Girard et le problème du mal*, edited by M. Deguy and J. Dupuy, 301–9. Paris: Grasset, 1982b.

Serres, M. *Détachement: Apologue*. Paris: Flammarion,1983a.

Serres, M. *Rome: Le livre des fondations*. Paris: Grasset, 1983b.

Serres, M. *The Five Senses: A Philosophy of Mingled Bodies*. London: Continuum, 1985 [2008].

Serres, M. *Statues: The Second Book of Foundations*. London: Bloomsbury, 1987 [2015].
Serres, M., edited by *A History of Scientific Thought: Elements of a History of Science*. Oxford: Blackwell, 1989 [1995].
Serres, M. *The Natural Contract*. Ann Arbor: University of Michigan Press, 1990 [2011].
Serres, M. *The Troubadour of Knowledge*. Michigan: University of Michigan Press, 1991a [1997].
Serres, M. 'Chapter III: Michel Serres'. In *French Philosophers in Conversation: Levinas, Schneider, Serres, Irigaray, Le Doeuff, Derrida*, edited by R. Mortley, 47–60. London: Routledge, 1991b.
Serres, M. *La légende des anges*. Paris: Flammarion, 1993a.
Serres, M. *Geometry: The Third Book of Foundations*. London: Bloomsbury, 1993b [2017].
Serres, M. *Atlas*. Paris: Flammarion, 1994.
Serres, M. 'Interview with Hari Kunzru'. (1995) Found at: https://www.harikunzru.com/michel-serres-interview-1995/
Serres, M. *Variations sur le corps*. Paris: Le Pommier, 1999.
Serres, M. *Hominescence*. London: Bloomsbury, 2001 [2019].
Serres, M. *The Incandescent*. London: Bloomsbury Academic, 2003a [2018].
Serres, M. 'The Science of Relations: An Interview'. *Angelaki* 8, no. 2 (2003b): 227–38.
Serres, M. *Branches: A Philosophy of Time, Event and Advent*. London: Bloomsbury, 2004 [2020].
Serres, M. *Récits d'humanisme*. Paris: Le Pommier, 2006.
Serres, M. *La Guerre mondiale*. Paris: Le Pommier, 2008a.
Serres, M. *Le mal propre: Polluer pour s'approprier?* Paris: Le Pommier, 2008b.
Serres, M. *Times of Crisis: What the Financial Crisis Revealed and How to Reinvent our Lives and Future*. New York: Bloomsbury, 2009 [2013].
Serres, M. *Biogée*. Paris: Le Pommier, 2010.
Serres, M. *Thumbelina: The Culture and Technology of Millennials*. Lanham, MD: Rowman and Littlefield, 2012 [2015].
Serres, M. *Pantopie ou le monde de Michel Serres, de Hermès à Petite Poucette*. Paris: Le Pommier, 2014.
Serres, M. *Le Gauchet boiteux: Puissance de la pensée*. Paris: Le Pommier, 2015.
Serres, M. *Darwin, Bonaparte et le Samaritain: Une philosophie de l'histoire*. Paris: Plon, 2016.
Serres, Michel. *Relier Le Relié*. Paris, Le Pommier, 2019. (for all following inquiries)
Serres, M., and B. Latour. *Conversations on Science, Culture and Time*. Ann Arbor: Michigan University Press, 1992 [1995].
Shapin, S. and S. Schaffer. *Leviathan and the Air-Pump: Hobbes, Boyle, and the Experimental Life*. Princeton: Princeton University Press, 1985.
Simons, M. 'The End and Rebirth of Nature? From Politics of Nature to Synthetic Biology'. *Philosophica* 47 (2016): 109–24.

Simons, M. 'The Many Encounters of Thomas Kuhn and French Epistemology'. *Studies in History and Philosophy of Science. Part A* 61 (2017): 41–50.
Simons, M. 'Obligation to Judge or Judging Obligations: The Integration of Philosophy and Science in French Philosophy of Science'. In *The Past, Present, and Future of Integrated History of Philosophy of Science*, edited by Emily Herring et al., 139–60 London: Routledge, 2019.
Simons, M., J. Rutgeerts, A. Masschelein, and P. Cortois. 'Introduction: Gaston Bachelard and Contemporary Philosophy'. *Parrhesia: A Journal of Critical Philosophy* 31 (2019): 1–16.
Smith, C. 'Translator's Introduction'. In *Atomistic Intuitions: An Essay on Classification*, edited by G. Bachelard, vii–xv. Albany: State University of New York Press, 2018.
Stark, R. 'Secularization, R.I.P'. *Sociology of Religion* 60 (1999): 249–73.
Steinle, F. 'Experiments in History and Philosophy of Science'. *Perspectives on Science* 10, no. 4 (2002): 408–32.
Stengers, I. *The Invention of Modern Science*. Minneapolis: University of Minnesota Press, 1993 [2000].
Stengers, I. *Power and Invention: Situating Science*. Minneapolis: University of Minnesota Press, 1997.
Stengers, I. 'The Cosmopolitical Proposal'. In *Making Things Public: Atmospheres of Democracy*, edited by B. Latour and P. Weibel, 994–1003. Karlsruhe: ZKM Center for Art and Media, 2005.
Stengers, I. *La vierge et le neutrino: les scientifiques dans la tourmente*. Paris: Seuil, 2006.
Stengers, I. *Cosmopolitics I*. Minneapolis: University of Minnesota Press, 2010.
Stengers, I. *Cosmopolitics II*. Minneapolis: University of Minnesota Press, 2011a.
Stengers, I. 'Wondering about Materialism'. In *The Speculative Turn: Continental Materialism and Realism*, edited by L. Bryant, N. Srnicek, and G. Harman, 368–80. Melbourne: re.press, 2011b.
Stengers, I. 'Introductory Notes on an Ecology of Practices'. *Cultural Studies Review* 11, no. 1 (2013): 183–96.
Stengers, I. *In Catastrophic Times: Resisting the Coming Barbarism*. Lüneburg: Open Humanities Press, 2015.
Stiegler, B. 'Exiting the Anthropocene'. *Multitudes* 60 (2015): 137–46.
Strum, S. and B. Latour. 'Redefining the Social Link: from Baboons to Humans'. *Social Science Information* 26, no. 4 (1987): 783–802.
Talcott, S. 'Errant Life, Molecular Biology, and Biopower: Canguilhem, Jacob, and Foucault'. *History and Philosophy of the Life Sciences* 36, no. 2 (2014): 254–79.
Tarot, C. *Le symbolique et le sacré. Théories de la religion*. Paris: La Découverte, 2008.
Taylor, C. 'Introduction'. In *The Disenchantment of The World: A Political History of Religion*, edited by Marcel Gauchet, ix–xv. Princeton: Princeton University Press, 1999.
Taylor, C. *A Secular Age*. Cambridge: Belknap Press, 2007.
Thom, R. 'Stop Chance! Silence Noise'. *SubStance* 12, no. 3 (1983): 11–21.

Tiles, M. 'Is Historical Epistemology Part of the 'Modernist Settlement'?' *Erkenntnis* 75, no. 3 (2011): 525–43.

Tirard, S. 'Monod, Althusser et le marxisme'. In *Une nouvelle connaissance du vivant: François Jacob, André Wolff et Jacques Monod*, edited by C. Debru, M. Morange and F. Worms, 75–88. Paris: Editions Rue d'Ulm, 2012.

Turchetto, M. 'Althusser and Monod: A 'New Alliance'?' *Historical Materialism* 17, no. 3 (2009): 61–79.

Vagelli, M. 'Historical Epistemology and the 'Marriage' between History and Philosophy of Science'. In *The Past, Present, and Future of Integrated History of Philosophy of Science*, edited by Emily Herring et al., 96–112. Routledge: London, 2019.

Watkin, C. 'Michel Serres' Great Story: From Biosemiotics to Econarratology'. *SubStance* 44, no. 3 (2015): 171–87.

Watkin, C. *French Philosophy Today: New Figures of the Human in Badiou, Meillassoux, Malabou, Serres and Latour*. Edinburgh: Edinburgh University Press, 2016.

Watkin, C. *Michel Serres. Figures of Thought*. Edinburgh: Edinburgh University Press, 2020.

Webb, D. 'The Virtue of Sensibility'. In *Michel Serres and the Crises of the Contemporary*, edited by R. Dolphijn, 11–29. London: Bloomsbury, 2018.

Weber, M. *The Protestant Ethic and The Spirit of Capitalism*. London: Unwin Paperbacks, 1985.

Weber, M. *From Max Weber: Essays in Sociology*. London: Routledge, 1991.

Williams, J. *Lyotard: Towards a Postmodern Philosophy*. Cambridge: Polity Press, 1998.

Woolgar, S. *Science: The Very Idea*. London: Tavistock, 1988.

Wunenburger, J., edited by *Gaston Bachelard: Science et poétique: une nouvelle éthique?* Paris: Hermann, 2013.

Wunenburger, J. *Gaston Bachelard, poétique des images*. Paris: Mimesis, 2012.

Zammito, J. *A Nice Derangement of Epistemes: Post-Positivism in the Study of Science from Quine to Latour*. Chicago: University of Chicago Press, 2004.

Index

Note: Page numbers followed by 'n' refer to notes.

Achever Clausewitz 160
acosmism 214, 215
Action Française 147
actor-network theory (ANT) 89, 96, 105, 107, 110, 112, 113
actors 7, 50, 106, 107, 109, 168, 186, 209, 213, 219
agriculture 84, 123, 125, 138, 145, 197, 215
Alain 72
alterity 178–81, 185, 193
Althusser, Louis 14, 36–48, 55, 77, 78
 science and ideology in 77–8
Althusserian legacy 47–8, 52, 63, 86
Althusserians 36–48
Anthropocene 12, 164, 197
 parliament of things and 197–220
anthropology, of science 65, 105, 141–69
archaic history 65, 141, 142
Assmann, Jan 191
Atlan, Henri 153, 154, 168
Atlas 147, 149
Attali, Jacques 153
attitude 18, 40, 48
autonomous self 69–72
axiomatics 133–5

Bachelard, Gaston 6, 9, 13–33, 37, 39, 48–50, 52–4, 61, 65, 68, 70, 73, 75–6, 81
 formation of scientific spirit 65–78
 Janus head of 52–6
 Michel Serres's critique of 19–33
 philosophy of science of 14–17
 and scientific self 68–76
Bachelard, Suzanne 29
Baillet, Jack 153
Baird, Davis 88
Barthes, Roland 2
Belaval, Yvon 29
Bensaude-Vincent, Bernadette 13

Bergson, Henri 7, 23, 75
biogea 190, 212, 217
Bloor, David 87, 182
Boissel, Jean-François 153
Bourdieu, Pierre 88, 89
bourgeois science 41
brewers, time 115–39
Brillouin, Léon 28, 89, 97–104
Brunschvicg, Léon 6, 15, 38, 53, 72–3
Buron, Robert 153
Butor, Michel 222 n.1

Caillois, Roger 146
Callon, Michel 89, 105–7
Canguilhem, Georges 29–31, 39, 40, 42, 45, 55–6, 62, 77, 81, 91, 142
capitalism 106, 133, 135, 137
Cassirer, Ernst 15
Cavaillès, Jean 62, 77
Chimisso, Cristina 66
circumstance 89, 103
Coblence, Françoise 153
communication 44, 45, 97, 101, 159, 168, 200–3, 219
Comte, Auguste 14, 22, 23, 27, 30–3, 46, 62, 174, 189
constructivism 19, 35, 51, 62, 89, 113, 189
 social constructionism 87–8, 198
cosmocracy 215, 217, 219
cosmopolitics 216
counterclaims 214, 216
counterproductivity 157
Crahay, Anne 1
credibility 89
current ecological crisis 8, 12, 199, 205

Dagognet, François 10, 87, 89–96, 109–11, 113
 and inscriptions 90–6

de Brosses, Charles   187
De la grammatologie   90
Deleuze, Gilles   7, 134-7, 219
De Rosnay, Joël   153, 154
Derrida, Jacques   2, 90, 91, 95, 100, 109, 222 n.1
Descartes   22, 23, 67
Despret, Vinciane   208
Détachement   3, 84
Dolphijn, Rick   4
Duby, Georges   146
Du culte des dieux fetishes   187
Dumézil, Georges   11, 142, 144-9, 152, 156, 169
  tripartite hypothesis   144-6
Dupuy, Jean-Pierre   11, 154-8, 161-3, 168
dynamic self   69-72

ecological crisis   164, 187, 210, 211
ecological objects   211
ecology   8-10, 12, 56, 138, 139, 165, 169, 199, 200
Écriture et iconographie   91
Einstein, Albert   37, 99, 155, 181
Éloge de l'objet   111
Entre le cristal et la fumée   154
epistemological break   35-59
  and Althusserians   36-48
epistemological profiles   76
epistemological rupture   9, 16, 21, 25, 37, 48, 54-5, 73
Eribon, Didier   147
Esthétiques sur Carpaccio   40
état naissant   70, 79
ethical substance   67, 75
Excluded third   1, 32, 84, 101, 108, 159, 165, 186, 201, 203, 204, 207, 211-12
Exo-Darwinism   123-4

Faurisson, Robert   120
fetish/fetishism   147-8, 187-93
Fichant, Michel   38
Foerster, Heinz von   154
Foucault, Michel   2, 10, 30, 61, 62, 66-8, 75, 77, 85, 146, 147, 160, 161, 165, 166
  care of the self   66-8

Frémont, Christiane   3, 13
French historical epistemology   6, 14
French object-oriented philosophy   87-113
Fuller, Steve   48, 174, 181

Gaia   164, 174, 187-93, 211, 212, 219
Galileo   37, 162, 172, 173, 185-7, 191, 192
Gattinara, Castelli   15
Gauchet, Marcel   11, 175, 177-81, 183-5, 192-5
  disenchantment of the world   178-9
  immanence and   180-1
general ecology   200
general theory of relations   1
Genèse   46, 147
Gingras, Yves   209
Girard, René   3, 11, 149, 155, 160, 162, 169
  and mimetic violence   149-52
Governance of Science   174
grande mystère   146
Great Story of the Universe   122-6
Greimas, Algirdas Julien   88, 91
Group of Ten (*Groupe des Dix*)   152-4
Guattari, Félix   134-5, 219

Hacking, Ian   36
Harman, Graham   87, 96
Hermes   1, 81-2, 158, 167
Hermès III, La traduction   106
Hermès series   9, 22, 29, 97, 100, 103, 116, 138
Hiroshima   20, 142-4, 159, 162, 172
historical epistemology   6, 30, 35, 37-8, 55, 61, 65
history of religion   178, 179
history of science   15, 17, 18, 30, 37, 38, 62, 63, 65, 66, 68, 69, 127, 143, 162
Hominescence   122
humanism   124-5, 134, 136-7
human sciences   64, 152, 155, 166
Hyppolite, Jean   29

idealism   30, 38, 43-4, 53-4, 63, 109, 112, 167
  linguistic idealism   88, 91-2

Idéologie et rationalité dans l'histoire des sciences de la vie   30, 55
ideology   32, 33, 39, 40, 47, 48, 50, 54, 55, 77, 78, 198
Iliad   146
Illich, Ivan   157
immanence   131, 134, 137, 179, 180–1, 190–2
immanentization   190–2
Indo-European languages   144, 145
information   23, 24, 93–5, 97–105, 206
    theory   7, 10, 20, 21, 27, 63, 91, 97–101
inscriptions   10, 89–96, 105, 106, 109, 110, 112
    materiality of   92–4
    paradoxical productivity of   95–6
    and translations   109–10
invention, pedagogy   4

Jacob, François   31, 41, 42, 100, 153
Jupiter   144–9
    and law   160–5
    and science   162–4

Kant, Immanuel   16, 118, 198, 213
Keeling, Charles   58
Keeling Curve   58
Klein, Ursula   92
Knorr-Cetina, Karin   89
Kojève, Alexandre   173
Kripke, Saul   121
Kurgan hypothesis   145

Laboratory Life   50, 90
Laborit, Henri   153
Lacan, Jacques   2
Lachelier, Jules   72–3
La condition postmoderne   116, 118, 119
La dialectique de la durée   75
Ladrière, Jean   1
La formation de l'esprit scientifique   31, 39, 61, 70
Lagneau, Jules   72–3
La Guerre Mondiale   149, 159
La Légende des Anges   1
La logique du vivant   100
La méthode   154
L'Anti-Œdipe   134

Latour, Bruno   2, 7, 10, 13, 35, 36, 56, 57, 80, 87–91, 93, 96, 97, 108, 110, 112, 134, 167, 173, 188, 192, 197, 198, 200, 216, 218, 222 n.2
    modernity and   126–30
Laurent, Alain   153
Le Cahier de l'Herne: Michel Serres   4
Le Cinq sens   3
L'économie et le vivant   154
L'économie libidinale   118, 134
Le contrat naturel   111, 164, 186, 200, 213
Lecourt, Domnique   30, 39, 44
Le Différend   120
Le Gaucher boiteux   5
Le hasard et la nécessité   41, 100, 104
Leibniz   1, 2, 20, 26, 97, 222 n.2
    new Leibnizianism   97
Le macroscope   154
L'enfer des choses   158
Le Parasite   108, 154
Leriche, René   81
Leroi-Gourhan, André   7, 153
Les cinq sens   51, 81, 111
Le Tiers-Instruit   80
Leviathan and the Air-Pump   112
Lévi-Strauss, Claude   31, 33
L'Incandescent   122
Linguistic turn   118, 198, 206, 208
    postlinguistic stance   199, 214
Lovelock, James   211
Lucretius   7, 25
Lyotard, Jean-François   10, 89, 116–22, 124, 128, 131–4, 136, 137, 139
    paganism   118–19
    postmodern fable   130–3

macrophilosophy   5–6
Mapping Michel Serres   4
Marcenay, Georges   147
Mars   141, 142, 144–9, 156–9, 167, 169
    and war   158–60
Marxism   33, 37, 40, 42–7, 118
    Serres's criticisms of   44–7
Materialism   43, 53–4, 75, 78, 104–5, 111–12, 208
    demanding materialism   208
    dialectical materialism   30, 43
    historical materialism   38, 121

intermaterialism 54
rational materialism 53
Maurras, Charles 147
Medical Nemesis 157
Merton, Robert K. 156
metanarratives 116, 118–21, 132–4
Michel Serres and the Crises of the Contemporary 4
microphilosophy 6
mimesis 149–50
mimetic violence 11, 149–52, 158–60, 164
   scapegoat mechanism 150–2
modernity 115–39
   postmodernity and 118–33
modern science 38, 67, 109, 173, 180, 183, 205, 211
molecular biology 97–100
Monod, Jacques 9, 40–4, 100, 104, 153
Morin, Edgar 153, 154
Mouffe, Chantal 161

Naming and Necessity 121
natural contract 8, 12, 165, 188, 198, 201, 213–15, 217, 219, 220
natural sciences 92, 152, 165, 174, 183
Nature 8, 11, 19, 26, 49, 69, 87, 88, 137, 165, 174, 179, 184–7, 190–2, 197–200, 203, 205, 206, 214, 216
   archetypes of 22
   end of 210–13
new informational ontology 100–5
new new scientific spirit 9, 21–7, 32, 33, 205
new scientific spirit 13, 18, 20, 33, 37, 52, 53, 142
noise 49, 80–2, 89, 101, 102, 104, 154, 168, 201–3, 207, 214, 215
Northwest Passage 152–6
'Noumène et microphysique' 49
Nous n'avons jamais été modernes 57, 89, 112, 192

Objectif 72 153
Order 20, 33, 46, 97, 101–5, 135, 153–5, 201–4, 207
   social 3, 8, 11, 46, 148–51, 153, 156, 157, 161, 162, 164

parasites 84, 85, 108, 200–3, 205–7, 211, 213, 214
   logic of 201–3
parliament of things 209–20
Passet, René 153, 154
passive objects 188, 203, 204, 208, 209, 215, 220
Pasteur, Louis 91, 107–9, 202
The Pasteurization of France 116
Paulson, William 2, 4
Pêcheux, Michel 38
Peden, Knox 77
Péguy, Charles 222 n.2
performativity 117, 120, 133–9
   criticizing 135–9
Petite Poucette 79
Petite réflexion sur le culte moderne des dieux faitiches 188
phenomenotechnique 9, 10, 23, 35–7, 48–50, 52–4, 57–9, 73
   in science studies 48–52
philosopher, new image of 27–8
Philosophie du non 17–19, 27, 76
philosophy of prepositions 1
Piette, Jacques 153
poetic self 74–5
political ecology 200, 211
political philosophy 28–33
Politiques de la nature 200
Posthumus, Stephanie 217
postmodernity 115, 118–33
presentism 18–19, 31, 68–9
Prigogine, Ilya 7, 20
proliferation 25, 26, 50–2, 54–9, 126, 127
   model of 25, 26, 50–2, 59, 199
Proto-Indo-European language 144, 145
psychoanalysis 61–4, 76
purification 10, 25, 26, 39, 50, 51, 53, 54, 56–9, 67, 73, 83, 84, 127, 137, 205, 209, 210
   chemical purification 53
   mental purification 54
   model of 25, 26, 31, 39, 47, 50–2, 57, 59, 199
   movements 183
   new model of 83–5
   practice of the self 61–86
   spiritual purification 31
   of subjects 127, 129

temporal purification   127
work of   57–8

quasi-objects   3, 8, 11, 12, 115, 156, 197–220
   counterclaims of   214–20
   ecology of   8, 10, 139, 200, 215
   negotiate with things   207–9
   omnipresence of   203–5
   parasite to symbiont   200–5
   scapegoat as   154–6
   witness   206–7
Quirinus   142, 144–9, 156–8, 167, 169
   and economy   157–8

Rameaux   122
rationalism   13, 15, 16, 24, 25, 30, 70, 71, 73, 74
Rationalisme appliqué   30, 71
ready-made science   56, 205
realism   48, 53, 64, 76, 87–8, 94, 110, 167, 183
   speculative realism   87–8, 198
Récits d'humanisme   122
rectification   54, 55
'Reform and the Seven Sins'   61–5
reification   89
relational ontology   10, 14, 51, 131, 133–9
relativism   24, 87
religion   11, 157, 162, 165, 169, 171, 173–9, 181, 183, 184, 189, 191–3
   contra-religion   191
Rématerialiser   111
representation   44, 95, 159–60, 166, 186, 194, 203, 205, 206, 207, 217
restricted ecology   200
Rheinberger, Hans-Jörg   36, 74
rhythmanalysis   75
Robin, Jacques   153
Rome   149
Rosenthal, Gérard   153
Rousseau, Jean-Jacques   154
Rousset, David   153

Santos, Lúcio Alberto Pinheiro dos   75
Sartre, Jean-Paul   89
Sauvan, Jacques   153
Schaffer, Simon   112

Schmidgen, Henning   96, 222 n.1
science
   bringing down to earth   182–4
   fetish to Gaia   187–93
   Jupiter and law   160–5
   laboratory and windmill   184–7
   Mars and war   158–60
   myth and critique   165–8
   Quirinus and economy   157–8
   secularization of   171–95
   sociology of   88, 89, 171
   and technology   7, 8, 35, 88, 111, 137, 152, 193, 198, 199
   and violence   156–68
science in the making   56, 57, 205, 215
scientific activity   17, 42, 104, 105
scientific body formation   78–86
   relational self   79–83
   self-forming activities of   85–6
scientific city   31, 73
scientific concepts   17, 69, 72, 76
scientific facts   58, 59, 88, 96, 102–4, 205, 206, 208
scientific knowledge   16, 25, 37, 38, 96, 104, 174, 182–4
scientific modernity   25
scientific practices   16–19, 24, 50, 51, 54, 56–9, 77, 94, 97, 174
scientific revolutions   15, 17, 18, 21, 64, 73
scientific self   10, 61, 68–76, 78, 79, 86
scientific spirit   16–18, 55, 62, 63, 65–78
scientific thought   18, 69, 72, 73, 76
A Secular Age   177
secularization   11, 171, 174–8, 180, 181, 183, 185, 190–3
   metaphors   11, 171, 173–5, 180, 182, 184, 190–3
   of science   171–95
   of the West   176–81
self   61, 65–73, 75, 79, 80, 83, 85
self-forming activities   67, 75, 85
sensitivity   8
Serres, Michel   3, 9, 40, 41, 89, 91, 97, 105, 106, 108, 112, 115, 116, 149, 154, 159, 186, 190, 199, 215, 222 n.2
Shapin, Steven   112
social activity   182, 183

social bond   139, 156
social relations   46, 73, 82, 83, 101, 147, 156
social sciences   82, 142, 148, 160, 165-7
social system   146, 168
sociology of scientific knowledge
  (SSK)   174, 175, 177, 181, 182
sociology of translation   89, 97, 105-7
solar systems   32, 122, 125
speech   90, 92, 133, 203
Spinoza   68, 77
Spinoza contra Phenomenology   77
Statues   111, 112, 163
Stengers, Isabelle   7, 20, 24, 35, 39, 56,
  188, 197, 198, 216, 219, 222 n.1
surrationalism   13-33
  and primacy of science   14-17
surveillance   165, 166
Symbiosis   200, 201, 213-14, 218-19

Taylor, Charles   177-8
technologies   12
  of control   58
  as negotiation   205-9
  of negotiation   58
  role of   9, 205, 206, 208
terrorism   160, 165
textism   88
thanatocracy   141-69
Thom, René   7, 42
totipotency   137, 139
transcendence   11, 137, 157, 171, 173,
  175, 179, 180-8, 192-3, 206, 216
  mini-transcendence   192-3
  pseudo-transcendence   157, 162-3
transcendental objects   111, 112
translation-inscription   93
translations   10, 57, 89, 95-111, 201
  information theory and molecular
    biology   97-100
  and inscription   109-10

new informational ontology   100-5
  process   106
  sociology of   105-9
*Treatise on the Improvement of
  Understanding*   68
TRF(H)   50, 103
tripartite hypothesis   144-6
  impact and criticism   146-7
tripartite system   145, 146, 169
truth   22, 25, 27, 29, 51, 52, 67, 68, 70,
  164, 166, 167, 180, 182, 184

unbillical thinking   5
universalism   124-5, 218

Variations sur le corps   79, 82
Vernant, Jean-Pierre   146
Verne, Jules   2
victim   120, 152, 155
violence   11, 26, 46, 47, 82, 83, 141-4,
  148-51, 158-63, 202
  and science   156-68
virtue of sensibility   8

warriors   145, 146, 151
water, air, fire, earth and life
  (WAFEL)   198, 217, 218, 220
Watkin, Christopher   2, 4, 5, 22, 26, 199,
  207
Webb, David   8, 9
Weber, Bernard   153
Weber, Max   176, 178, 179, 181
Witness   59, 122, 161, 206-9, 210
Wittgenstein, Ludwig   118
Woolgar, Steve   88, 90-1
word-for-word translation   107
workers   120, 121, 146
world-objects   138, 143
World Wars   77, 119, 144, 145, 148, 160,
  164

www.ingramcontent.com/pod-product-compliance
Lightning Source LLC
Chambersburg PA
CBHW062139300426
44115CB00012BA/1981